Drones in IoT-enabled Spaces

Drones in IoT-enabled Spaces

Fadi Al-Turjman

CRC Press
Taylor & Francis Group
Boca Raton London New York

CRC Press is an imprint of the
Taylor & Francis Group, an **informa** business

CRC Press
Taylor & Francis Group
6000 Broken Sound Parkway NW, Suite 300
Boca Raton, FL 33487-2742

© 2019 by Taylor & Francis Group, LLC
CRC Press is an imprint of Taylor & Francis Group, an Informa business

Library of Congress Cataloging-in-Publication Data

Names: Al-Turjman, Fadi, editor.
Title: Drones in IoT-enabled spaces / [edited by] Fadi Al-Turjman.
Description: Boca Raton, FL : CRC Press/Taylor & Francis Group, 2019. |
Includes bibliographical references and index.
Identifiers: LCCN 2019013224 (print) | LCCN 2019013980 (ebook) |
ISBN 9780429294327 (e) | ISBN 9780367266387 (hb : acid-free paper)
Subjects: LCSH: Drone aircraft—Control systems. | Drone aircraft—Industrial applications. | Internet of things.
Classification: LCC TL589.4 (ebook) | LCC TL589.4 .D75 2019 (print) |
DDC 629.133/3902854678—dc23
LC record available at https://lccn.loc.gov/2019013224

Visit the Taylor & Francis Web site at
http://www.taylorandfrancis.com

and the CRC Press Web site at
http://www.crcpress.com

Printed and bound in Great Britain by
TJ International Ltd, Padstow, Cornwall

Sometimes you can find words to fill in a 250 pages' book, but you can't find a word to thank somebody without whom the book itself wouldn't be realized... Thanks Sinem.

Thanks to my parents, my brother, my sisters, and my kids... Thanks to all who standby...

Fadi Al-Turjman

"Great things in business are never done by one person. They're done by a team of people."

Steve Jobs

Contents

Author

Prof. Fadi Al-Turjman received his Ph.D. degree in computer science from Queen's University, Canada, in 2011. He is a professor at Antalya Bilim University, Turkey. He is a leading authority in the areas of smart/cognitive, wireless and mobile networks' architectures, protocols, deployments, and performance evaluation. His record spans more than 200 publications in journals, conferences, patents, books, and book chapters, in addition to numerous keynotes and plenary talks at flagship venues. He has authored/edited more than 12 published books about cognition, security, and wireless sensor networks' deployments in smart environments with Taylor & Francis and Springer (top-tier publishers in the area). He was a recipient of several recognitions and best paper awards at top international conferences. He also received the prestigious *Best Research Paper Award* from Elsevier *COMCOM Journal* for the last 3 years prior to 2019, in addition to the *Top Researcher Award* for 2018 at Antalya Bilim University, Turkey. He led a number of international symposia and workshops in flagship IEEE ComSoc conferences. He is serving as the lead guest editor in several journals, including the *IET Wireless Sensor Systems, Springer EURASIP, MDPI Sensors, Wiley & Hindawi WCM, Elsevier Internet of Things,* and *Elsevier Computer Communications.*

Chapter 1

UAVs in Intelligent IoT-Cloud Spaces

Fadi Al-Turjman
Antalya Bilim University

Enver Ever and Murat Fahrioglu
Middle East Technical University

The wireless cellular communications infrastructure mainly depends on base station systems (BSS) that are responsible for ensuring communications of associated nodes and user equipment (UE). Under normal circumstances, the cellular and infrastructure-based systems work effectively. However, in events of unexpected conditions and natural disasters, such systems are relatively fragile and can easily be disrupted. During a natural calamity, the wireless communications infrastructure can be severely affected, where one or more BSS can stop working. The disruption in the operation of BSS affects the communications of interconnected devices. In such circumstances, flying ad hoc networks can assist as a substitute to provide structureless communications framework for communicating emergency and safety information using unmanned aerial vehicles (UAVs).

Recent developments in microelectromechanical systems (MEMS) technology and very large-scale integration have been influential in transforming large BSS to minute structures, which enables the adaptation of small-sized drones (or UAVs). UAVs are capable of the replicating technology features of BSS and can be used to form a small coverage area. UAVs, with the ability to move autonomously and to hover over the affected area, can function as a small cell to establish communications with the active UE in the designated emergency coverage area. Hypothetically, with the presence of sufficient UAVs, the communications outage

1

area in vulnerable regions can be fully covered. The restoration of a communication network in such areas using UAVs provides a rapid and reliable alternative to reconfigure and replicate necessary functionalities of the affected BSS. These drone small cells (DSCs) can also be used to enhance and extend communication coverage in disaster areas where on-ground repairs are not feasible. The ability of DSCs to reposition itself and respond to the UE by reducing distance extends coverage, decreases outage probability of the UE in coverage zones, improves bandwidth efficiency, and optimizes system throughput.

However, due to the nature of sensitivity of such situations, additional constraints such as delay and reliability are required, which are very challenging. Moreover, the incorporation of appropriate information-based urgency index in ad hoc networks is also very important. In fact, communications in emergency networks can be classified into a number of precedence levels, where alerting messages, well-being messages, control messages, distress calls, and data collection schedules can be characterized separately to optimize the ongoing communications between UAVs. Therefore, a suitable *intelligent* mechanism is needed to associate priority levels with these calls, messages, and schedules. Providing multihop collaboration among UAVs in an attempt to reach possible urgent services that can be provided by the cloud facilities, where machine learning (ML)-based approach is employed for the adaptation of existing configuration can significantly improve the services that DSCs can provide. Automating the collection and analysis of data has the potential to lead to more robust and intelligent systems that can save lives and time for the emergency and rescue teams involved.

1.1 Intelligence in UAVs

Recently, artificial intelligence, specifically ML, showed an outstanding performance in complicated tasks that require human-like intelligence and intuition to perform. ML is suited for the situations where there are no defined rules for performing a task, and instead, the rules are learned from real data. ML is capable of detecting hidden structures in the data to make smart decisions. ML techniques can be classified in general into three main categories. This classification is mainly based on the kind of data and the objective of the task. The three categories are as follows.

1. Supervised learning: This is the well-established and most used technique. Supervised learning techniques use data to make accurate predictions and learn the mapping between the input and its corresponding output while receiving a feedback during the learning process to identify things based on similar features. Approaches in this category are used to predict an outcome or the future or to classify the input to a set of desired classes. Most common approaches in this category can be regression algorithms, support vector

machine, and neural network approaches. In order to introduce the training employed in these techniques, usually a function (linear, nonlinear, polynomial, fully connected neural network, etc.) that can best approximate the relation between the input and output data is defined. Then, a cost function is set to tell the learner how much it is far from the best answer, so it acts as a feedback signal. In turn, this signal is used to update the parameters of the function at each iteration. At the end, this function is used to make the prediction of future input or classify unseen data.

2. Unsupervised learning: Unlike supervised learning that uses labeled data, unsupervised learning has no labels and no feedback signal. This technique is mostly used to find the hidden structure of the data and move it into similar groups. So, they are mainly used for pattern detection and descriptive modeling. These types of algorithms are promising to achieve general artificial intelligence, but they usually lack behind supervised learning in terms of accuracy and computation time. K-means and autoencoder are the most known unsupervised algorithms.

3. Reinforcement learning (semisupervised): This technique resembles to highly extend the way humans learn and navigate through their daily life tasks. Reinforcement learning is neither fully supervised nor unsupervised, but it's a kind of hybrid approach.

Appling any of these ML techniques in a DSC-based coverage network can restore the necessary links in the communications outage area while ensuring minimal delay for emergency communications and maximum network throughput for better bandwidth/resource utilization. Further improvements in ML techniques design for infrastructureless UAV-based communications in emergency personal sensor networks (PSNs) can also support in disaster communications, using new technologies such as device-to-device (D2D), machine-to-machine, internet of things (IoT) communications. For example, authors in Refs. [1,2] examined how the in-coverage UE deliver the elementary network services to out-of-coverage UE by relaying their data to eNB (evolved NodeB) as base station. The study investigated the selection of an in-coverage UE in PSN. The findings suggested that there is no centralized entity in PSN to assist the discovery and synchronization of UE and should separately be addressed, which results in high energy consumption and delay. In addition, authors in Ref. [3] outlined that UE selection process was also highly critical because both in- and out-of-coverage UE have very limited energy and processing capability. There was limited reliability in terms of availability, throughput, and traffic handling capabilities of UE and cannot concurrently handle PSN demands. Therefore, the use of DSCs is well suited for PSNs. The suitability of DSCs in PSNs is primarily attributed to self-organization, mobility, and delay minimization abilities of DSCs.

In Refs. [4,5], UAVs are proposed as a part of a system targeting postdisaster scenarios. The subsystems running on each UAV are explained and evaluated using a prototype helicopter to prove the efficiency of the navigation subsystem. The long-term evolution-unlicensed (LTE-U) technology is proposed in Ref. [6] for DSCs to enhance the achievable broadband throughput for postdisaster assistance. An ON/OFF game-based mechanism is employed for effective use of LTE-U, and to reach a correlated equilibrium. Numerical simulations are employed in Ref. [7] to study the coverage that can be provided by UAV-based base stations. The study attempts to minimize the number of stops and amount of delays for a single UAV that needs to visit various positions to completely cover the potential disaster area. This study is further extended in Ref. [8] for multiple UAVs. A framework is proposed for optimizing the 3D placement and the mobility of UAVs. Simulations performed using MATLAB® provide results that show significant enhancements using the proposed approach, especially in terms of reductions in transmission power of IoT devices and system reliability. Through these results, the significance of intelligent decisions in terms of UAV deployment and repositioning has been emphasized.

1.2 Collaborative UAVs in Cloud

The decision-making and evaluation processes of cloud-based studies in this area are mainly dependent on high-level analytical abstractions of scenarios considered. We believe that there are factors above the physical and data link layers that can affect the optimization of heterogeneous infrastructures that can involve conventional base stations, D2D communications of UEs, and DSCs. For incorporating the potential complexities of more realistic scenarios, it is possible to provide communication between UAVs and the existing cloud facilities to use more sophisticated approaches such as ML for the analysis [9].

In Ref. [10], the authors propose a framework to use UAV support for wireless powered communication (WPC) techniques that mainly focus on providing energy to the UEs of potential victims in disaster areas. The mobility features of UAVs are employed to improve the conventional WPC techniques that are mainly dependent on a static access point responsible for charging a set of wireless nodes in the downlink. A distributed resource management mechanism is proposed in this study to optimize the public safety IoT (PS-IoT) devices' uplink transmission powers and UAV positioning. However, considering allocation of uplink and downlink resources and optimization using various methods based on game theory may not be sufficient, since higher level of simulations where traffic conditions, mobility-related issues, and availability of other facilities should also be considered together with facilities provided by UAVs. Furthermore, considering the limited flying time mainly due to the limited energy resources of UAVs, the optimum configuration for the transmission of safety critical information becomes even more critical.

A drone cooperation scenario is considered in Ref. [11]. The UAV-based base stations are employed together with conventional base stations in an attempt to aid the disaster-struck regions where terrestrial infrastructure is damaged. The main focus of this study is efficient power allocation strategies for the microwave base station as well as smaller UAV-based base stations. The power control strategy presented is self-adaptive depending on the interference threshold employed as well as data rate requirements. Factors such as UAV altitude and number of ground users are considered with an analytical abstraction for simulations. The importance of incorporation of UAVs in the multitier heterogeneous networks for better network coverage and capacity is emphasized in this study as well [12].

1.3 Conclusion

Research in DSC is still in its infancy, and many practitioners and academics are keen to pursue their research in this scholarly area. The use of DSC-based solutions, where an infrastructure can be made available very rapidly, particularly, for emergency communications in disaster-affected areas, is a very promising solution.

The research work on this topic mainly advocates the following reasons for the employment of DSC-based solutions in PSNs: (1) UAVs are able to hover at higher altitude to provide a suitable height gain; (2) through energy sustainability, UAVs can be made suitable for PSNs, since the main aim is to exchange emergency-related information for short durations; (3) while hovering, UAVs improve connection reliability and offer better connectivity and efficiency for UE; (4) the usage of DSCs can allow efficient use of bandwidth and improve frequency reusability; and (5) the use of DSCs will result in rapid deployment of communication network in disaster-affected areas where early involvement is essential. The utilization of DSCs in critical scenarios has the potential of introducing significant advantages, since due to their mobility, flexibility, and adaptability, the DSCs are able to provide coverage and capacity exactly where and when it is needed even under such circumstances that other means of communication services are not available.

The main areas of interest that requires improvements for development of DSCs are as follows: (1) optimized on-demand communications should come with enhanced throughput to support highly resilient networks within critical and emergency scenarios; (2) ad hoc on-demand formation of small cells should support enhancement of the number of users to be served by and at the same time prioritize the communications of rescue workers and first responders, reporting from the disaster-affected areas; (3) A priority-wise channel access establishment should also be provided for emergency-related communications, which reduces channel access delay within DSCs; (4) the deployment, mobility, and coverage-based issues, such as potential areas with higher numbers of victims, should be addressed.

References

1. K. Ali, H. X. Nguyen, P. Shah, Q. T. Vien, and N. Bhuvanasundaram, Architecture for public safety network using D2D communication, in *2016 IEEE Wireless Communications and Networking Conference*, Doha, Qatar, April 2016, pp. 1–6.
2. K. Ali, H. X. Nguyen, P. Shah, Q. T. Vien, and E. Ever, D2D multi-hop relaying services towards disaster communication system, in *2017 24th International Conference on Telecommunications (ICT)*, Limassol, Cyprus, May 2017, pp. 1–5.
3. K. Ali, H. X. Nguyen, Q. T. Vien, P. Shah, and Z. Chu, Disaster management using D2D communication with power transfer and clustering techniques, *IEEE Access*, vol. PP, no. 99, 1, 2018.
4. O. Oubbati, A. Lakas, F. Zhou, M. Güneş, and M. Yagoubi, A survey on position-based routing protocols for Flying Ad hoc Networks (FANETs), *Vehicular Communications*, vol. 10, 29–56, 2017.
5. G. Tuna, B. Nefzi, and G. Conte, Unmanned aerial vehicle-aided communications system for disaster recovery, *Journal of Network and Computer Applications*, vol. 41, 27–36, 2014.
6. A. Dasun, I. Guvenc, W. Saad, and M. Bennis, Regret based learning for UAV assisted LTE-U/WiFi public safety networks, in *Global Communications Conference (GLOBECOM), 2016 IEEE*, Washington, DC, 2016, pp. 1–7.
7. M. Mozaffari, W. Saad, M. Bennis, and M. Debbah, Unmanned aerial vehicle with underlaid device-to-device communications: Performance and tradeoffs, *IEEE Transactions on Wireless Communications*, vol. 15, no. 6, 3949–3963, 2016.
8. M. Mozaffari, W. Saad, M. Bennis, and M. Debbah, Mobile unmanned aerial vehicles (UAVs) for energy-efficient internet of things communications, *IEEE Transactions on Wireless Communications*, vol. 16, no. 11, 7574–7589, 2017.
9. F. Al-Turjman, M. Z. Hasan, and H. Al-Rizzo, Task scheduling in cloud-based survivability applications using swarm optimization in IoT, *Transactions on Emerging Telecommunications*, 2018. doi:10.1002/ett.3539.
10. S. Dimitrios, E. Tsiropoulou, M. Devetsikiotis, and S. Papavassiliou, Wireless powered public safety IoT: A UAV-assisted adaptive-learning approach towards energy efficiency, *Journal of Network and Computer Applications*, vol. 123, 69–79, 2018.
11. S. Raza, S. A. Hassan, H. Pervaiz, and Q. Ni, Drone-aided communication as a key enabler for 5G and resilient public safety networks, *IEEE Communications Magazine*, vol. 56, no. 1, 36–42, 2018.
12. F. Al-Turjman, Cognitive routing protocol for disaster-inspired Internet of Things, *Elsevier Future Generation Computer Systems*, vol. 92, 1103–1115, 2019.

Chapter 2

Deployment Strategies for Drones in the IoT Era: A Survey

Fadi Al-Turjman
Antalya Bilim University

Mohammad Abujubbeh and Arman Malekloo
Middle East Technical University

2.1 Introduction

On one hand, the advancement in internet of things (IoT) technology enabled connectivity to a large number of smart devices that can be accessed at any time, from everywhere, using anything [1]. On the other hand, Drone technology, known as unmanned aerial vehicle (UAV), witnessed a vast attention in the recent years due to the advantages they can offer and their deployment flexibility. To this extent, both technologies form a promising paradigm that offers a wide range of applications in the modern smart city (SC) known as internet of Drones [2]. Consequently, Drones, as shown in Figure 2.1, can be used in public safety, area coverage, short-term large-scale emergency events [3], vehicle tracking and congestion management [4], coastal investigations [5], and forestry applications [6], to name a few. However, optimal deployment of Drones is considered as one of the biggest challenges that Drone technology faces [7]. With proper planning of Drones' positioning (deployment) in the 3D space, one can greatly benefit from their unique features, such as network reliability enhancement through area coverage schemes

7

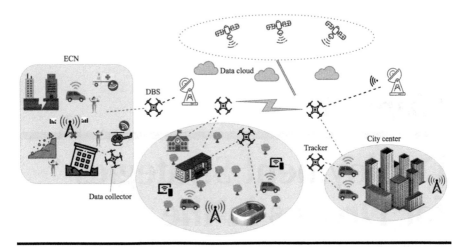

Figure 2.1   Drone deployment architecture and applications.

in Emergency Communication Networks (ECNs). Drone optimal deployment will not only enable major cost reductions such as minimizing number of Drones, for instance, taking into account a minimum Quality of Service (QoS) provision, but also provide secure end-to-end authentication process that mitigates vulnerability to threats and cyberattacks. Further benefits such as minimized energy consumption in the network and enhanced network connectivity are also accomplished through optimal Drone deployment in the desired network.

2.1.1 Scope of This Survey

Since deployment of Drone-based systems is critical, it is important to identify their desired features and specify existing placement strategies. This survey starts with the static deployment classification to have a better vision for strategic investigations in stationary placement methods, their objectives, and possible role-based planning. Accordingly, we propose a two-fold classification: (1) Controlled Drone deployments and (2) Random Drone deployments, while satisfying primary objectives and considering varying roles of the Drone. We believe that Drones' deployments are different from other communication systems' deployment due to their dynamic topologies and newly emerging communication protocols, standards, and technologies. Furthermore, variety of role-based deployment aspects are encountered during the implementation of various use cases mentioned in this survey. In comparison to other existing communication networks deployments, Drones are believed to be an adequate alternative because they do not require an active pilot [8]; instead, they can be remotely controlled, which is a great advantage, especially, in dangerous locations. Furthermore, in a wireless sensor network (WSN), sensor nodes are known for their energy and communication range constrains, which means that the direct transmission of data packets from sources to destinations is difficult. Thus, it is worth to employ off-the-shelf mobile Drones to gather data packets from the source and deliver them to the desired destination [9].

Putting the previous notes into consideration, we intend to review Drone deployment strategies in a synergistic manner by drawing on existing related literature efforts.

2.1.2 State of Surveys

There have been a few attempts in the literature towards providing a survey about deployment aspects of the Drone technology. For example, in Ref. [10], authors provide an insight on the technological improvements and future direction of autonomous Drones by reviewing Drone design and manufacturing challenges as well as their associated regulative issues. In Ref. [11], a review on Drone-based businesses in Chiba and challenges related to research and ImPACT (immediate post-concussion assessment and cognitive testing) program throughout the entire country, Japan. In line with IoT advancements, authors in Ref. [12] comprehensively review Drone technology by discussing Drone use cases and services, physical collision issues and avoidance methods, enabling communication technologies, data gathering, forwarding, and processing aspects in Drone technology, and future research suggestions accordingly. On the other hand, authors in Ref. [13] discuss Drone-based forestry applications by analyzing the ongoing research efforts. Moreover, authors in Ref. [14] explain Drone physical configuration and structure with a discussion on their relocation control mechanisms and provide research directions accordingly. Ref. [15] provides a thorough survey on Drones, including a classification metric according to Drone types and their respective applications, design challenges related to microDrones as well as approaches for increasing their endurance, and limitations and future solutions for Drone implementations. Furthermore, Ref. [16] investigates the advancement of space exploration-based Drone deployment, considering design issues associated with their fabrication with future research suggestions accordingly. Ref. [17] presents a comprehensive survey that provides an in-depth understanding of Drone characteristics and their applications, classifies Drone operation modes (such as area coverage, search operations, data gathering in WSNs, and communication link allocations) as well as integration of Drones with other types of vehicles, and discuss existing challenges related to Drone deployments with future research suggestions. In addition, radar-based Drone monitoring techniques are reviewed in Ref. [18] by reporting literature works surrounding Drone detection and classification approaches as well as providing future research recommendations. While Ref. [19] reviews the ongoing research attempts related to Drone deployment for building inspection purposes, with a case study validating the proposed framework on Syracuse University campus to demonstrate the procedure. Finally, authors in Ref. [20] review integrating Drone technology into medicine and showcase specific applications that, with the

Table 2.1 Comparison of Related Surveys

Ref.	Year	Brief Description
[20]	2018	Drone-based medicine applications and healthcare access enhancement
[19]	2018	Drone-based building inspection and case study conducted on Syracuse University campus
[18]	2018	Drone detection and classification approaches and open research issues
[17]	2018	Drone characteristics, applications, Drone operation mode, Drone-vehicle integration, deployment challenges, and future research suggestions
[16]	2018	Space-exploration Drones, design and fabrication issues, and open research issues
[15]	2017	Drone type classification, applications, challenges of microDrones, Drone endurance, and open research issues
[14]	2017	Drone physical configuration, relocation control mechanisms, and open research issues
[12]	2016	Drone use cases, physical collision issues, and avoidance methods, enabling communication technologies, data gathering, forwarding, and processing aspects in Drone technology and open research issues
[11]	2016	Drone-related businesses in Chiba and challenges related to research and ImPACT innovation program throughout Japan
[13]	2015	Drone-based forestry applications
[10]	2015	Drone design, manufacturing challenges, as and the associated regulative issues

presence of Drones, are believed to enhance healthcare access, including prehospital emergency, accelerating laboratory diagnostic testing, and surveillance tasks. The aforementioned surveys are listed in Table 2.1 according to their publication year.

To the best of our knowledge, none of the existing literature surveys discusses Drone technology implementation strategies, taking into account deployment and objectives, and application areas. Therefore, we outline our contributions to literature as follows, this survey

■ Provides a classification of Drone deployment approaches; static vs. dynamic deployments to have better understanding on the characteristics of each type by analyzing the ongoing research efforts.

- Discusses static Drone deployment methodologies (random and controlled), deployment objectives, and role-based Drone placements.
- Classifies dynamic Drone deployment into scheduled and on-demand approaches as introduced into literature and outlines the associated relocation issues.
- Explains selected Drone technology applications in the context of IoT and SCs and outlines selected Drone brands existing in the market as well as their deployment issues.
- Overviews open research issues in Drone deployments that future research may consider.

2.2 Static Positioning of Drones

Recently, Drones gain an increased attention due to the advantages they can offer in a wide range of applications. Static Drone deployment implies that a Drone or a set of Drones are deployed in the desired area to handle a specific task. In this context, Drones can hover above a certain location to provide accessibility and real-time monitoring schemes or a relay node as a backup solution for connection loss in communication networks. Static Drone deployment may further be categorized into controlled vs. random deployment depending on the application and objective. Therefore, in the following subsections, we intend to present the ongoing research efforts targeting Drone static deployment methodologies. We also categorize literature works according to their different objectives.

2.2.1 Deployment Methodology

In Drone static deployment, two main methodologies are used for the deployment: controlled and random deployments.

2.2.1.1 Random Drone Deployment

In this method, Drones are deployed over the desired region in a random manner that does not follow a certain arrangement. It can also be further classified into weighted random vs. unweighted random deployments as illustrated in Figure 2.2a and b. Unlike unweighted-random deployment, weighted-random deployment considers a certain factor in the deployment randomness. For example, in a disaster scenario, more Drones are required to hover close to the event center above the location while relatively less Drones are deployed outside the event zone. This type of deployment can be used mostly in temporary applications such as disaster rescue and surveillance, where the maximum coverage of the location is the priority regardless of Drone number or energy consumption. The unweighted approach, however, can be used in large-scale monitoring or search applications in which the

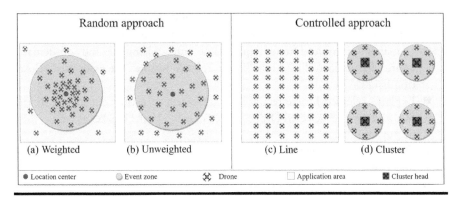

Figure 2.2   Drone deployment methodologies: (a) weighted random, (b) unweighted random, (c) line-based, and (d) cluster-based.

deployment is not constrained to location, cost, or energy consumption such as forestry monitoring and wilderness search and rescue applications.

2.2.1.2 Controlled Drone Deployment

Unlike random deployment, this approach follows a deterministic arrangement where the Drones are deployed in a line or cluster approach as shown in Figure 2.2c and d, respectively, depending on the application and arrangement of ground users or nodes that can also be deployed in a random or uniform manner [21,22]. Controlled deployment is essential for applications that require a certain arrangement of Drones. For instance, tracking and localization of vehicles passing through a highway may require a controlled line topology. Similar to Ref. [23], the authors attempt to position Drones on overhead transmission lines in order to report the status of transmission lines and track passing vehicles where the global navigation satellite system is not available. On the other hand, base stations (BSs) deployment can follow a cluster deployment topology having one special Drone called cluster head (CH), responsible for data transmission to data banks or clouds.

Deployment methodology is an essential parameter that requires careful consideration in Drone deployments. Future research may consider deployment topology as to enhance network performance. Similar to IoT-inspired WSNs, Drone-based deployment methodology affects different network parameters.

2.2.2 Role-Based Placement Strategies

Controlled static deployment can be further categorized based on the role of Drone. First, in wireless networks, Drones can be used as relay nodes that connect data sources with destinations. This type of Drone deployment can be represented as base stations or access points that help increasing the network reliability by providing alternative data routes in emergency cases where data transmission routes are lost due to different reasons such as link failure or increased data congestions. Second, Drones can be deployed as data collectors to report real-time information in a certain application. Usually, this type of deployment is greatly useful in applications characterized with environmental-harsh conditions or tracking and localization applications. This includes applications such as disaster management and rescue, military services, and structural health monitoring schemes. In the following subsections, we present the aforementioned role-based deployment strategies, and Table 2.2 categorizes selected research efforts that target static Drone deployment accordingly.

2.2.2.1 Relay Drone Placement

In the rich body of literature, many research efforts have proposed the usage of Drones as relays in communication networks. Considering the mobility and deployment flexibility of Drones, they can be deployed to cover an area of interest and hover while providing a connection between ground facilities and utilities. In line with IoT paradigm advancement, Drones can be employed in various networking technologies such as cellular and next-generation 5G networks and WSNs. Multiple Drones may also be used in ECNs to restore the communication link with a certain area during disaster events [24–26], as illustrated in Figure 2.1. In the light of Drone deployment as relay nodes, authors in studies [27,28] consider determining the number of Drone base station (DBS). In Ref. [27], authors target in determining the number of Drones while ensuring the required QoS level for a Marco-urban application scenario that operates at a radio range of 2 GHz [29], including a uniformly distributed 2,000 users. In Ref. [28], authors attempt to design a UAV for large-scale disaster scenarios. They investigate the effects of Drone height and user-service reduction on the necessary number of Drones to provide the necessary coverage over randomly deployed multiple users. The used Drone operates at a radio range of 5 MHz. Furthermore, Ref. [30] proposes a 2.5-GHz-based DBS model that considers path loss reduction at the cost of number of covered users. In contrary, Refs. [21,31,32] consider coverage as their objective. A bare bones fireworks algorithm is employed in Ref. [21] that targets DBS coverage range with minimum number of Drones used. Unlike Ref. [32] that studies coverage probability for a single user at a carrier frequency of 1,800 MHz, Ref. [31] considers multiple users and considers intercell signal interference as their constraint. In line with 5G communication technology advancements, authors in Ref. [33] propose a Drone-energy efficient cell outage compensation approach that ensures minimum QoS to users. Similarly, Ref. [34] considers 5G networking paradigm; however, it examines the human blockage effects on Drone design parameters such as coverage and number of users served, where they consider a maximum number of 100 randomly distributed users. Authors in Ref. [35] target coverage range as well consider network efficiency as a secondary objective as a response to Drone location variation.

Table 2.2 Drone Static Deployment

Role	Application Area	Ref.	Objective(s)	Radio Ranges	Ground User/Node		Connectivity Goals/ Constraints
					Number	Topology	
Relay node	Cellular networks	[27]	Number of Drones	2 GHz	2,000	Uniform	QoS
		[31]	Coverage	-	Multiple	Random	Intercell interference
	ECNs	[28]	Number of Drones	2.5 GHz	224	Random	Drone height and coverage
		[32]	Coverage	1,800 MHz	Single	-	-
	WSNs	[30]	Path loss minimization	2.5 GHz	50	Random	Number of covered users
		[21]	Coverage	-	30	Random and uniform	Number of Drones
		[36]	Fairness performance	2 GHz	100	Random and uniform	Number of Drones
		[37]	Energy efficiency	700 MHz	100	Random and uniform	Minimum transmit power
	5G networks	[33]	Energy efficiency	2.1 GHz	10	Random	QoS
		[34]	Coverage and number of users	28 GHz	100	Random	Human blockage

(Continued)

Table 2.2 (Continued)   Drone Static Deployment

Role	Application Area	Ref.	Objective(s)	Radio Ranges	Ground User/Node		Connectivity Goals/ Constraints
					Number	Topology	
Data collector	Environment-harsh conditions	[39]	Desert-surface albedo measurement	434 MHz	6	Uniform	-
		[40]	Luci eruptive site monitoring	2.4 GHz	Single	-	-
		[41]	Acid mine drainage monitoring	-	Single	-	-
	Tracking and localization	[42]	AUAV tracking	-	-	Random	-
		[43]	Single-shot vehicle detection	-	Single	Random	Drone resource constraint
		[44]	Range-free sensor node localization	-	50, 100, and 250	Uniform	-
		[22]	Target monitoring	-	30	Random and uniform	Number of Drones

In Ref. [36], authors target network fairness performance, taking into account the number of Drones at a carrier frequency of 2 GHz considering two scenarios: random and uniform arrangements of 100 ground users. Similarly, Ref. [37] uses the same user arrangement scenarios however, and considers a Drone energy efficiency with a minimum required transmission power at a frequency of 700 MHz.

2.2.2.2 Placement of Data Collectors

Drones can be deployed as data collectors in various applications such as providing access to harsh environments and tracking objects or wilderness search and rescue schemes [38]. In both scenarios, Drones provide real-time monitoring in the area of interest. For instance, Refs. [39–41] target designing Drones for providing accessibility to specific harsh-conditioned geographical locations. In Ref. [39], authors propose an albedometer and a Drone for evaluating the surface reflectance in the Black Rock Desert, Nevada. Ref. [40] designs the Luci Drone that enables monitoring the Luci mud eruption in NE Java Island, Indonesia. In contrary, Ref. [41] proposes a Drone that targets acid mine drainage monitoring in the Sokolov Lignite district. Refs. [22,42–44] target tracking and localization applications. Ref. [42] proposes an algorithm for accurately localizing amateur UAVs (AUAVs) using surveillance UAVs, considering computational efficiency enhancement. Authors in Ref. [43] provide a convolutional neural network-based algorithm for real-time vehicle detection system, taking into account localization accuracy. On the other hand, Ref. [44] attempts to exploit the advantages of Drones instead of anchor nodes for sensor node localization in highly dense networks. The study is conducted on a uniform arrangement of 50, 100, and 250 sensor nodes in the network. Study [22], in contrary, uses elephant herding algorithm to monitor 30 targets, considering both random and uniform arrangements aiming at coverage of all targets and the least number of deployed Drones.

In the light of IoT advancements, facilitation of further tracking and localization applications is necessary. Arrangement of Drone nodes in the network is also an essential parameter depending on the application topology.

2.2.3 Primary Objectives for Deployment

Drone deployment heavily relies on application topology and requirements. The ultimate goal is to achieve an ideal Drone-based network that offers an enhanced performance for a certain parameter. However, ideality differs from one application to another. For example, maximizing range of coverage in a certain application would require large number of Drones, which may result in increased costs and data congestions. Therefore, in the following subsections, we intend to present the essential deployment objectives.

2.2.3.1 Area Coverage

Area coverage using Drones has gained large attention in the recent years. Authors in literature propose different coverage schemes based on different metrics and application topologies. In Ref. [38] for example, authors present a method using UAV to provide video coverage for wilderness search-and-rescue applications. For assessing the model efficiency, they consider different metrics such as image resolution, number of observations, and viewing angle variations. On the other hand, Ref. [45] targets providing maximal radio coverage, considering the placement altitude as a function of maximum allowed path loss.

Design metrics in this objective requires further attention so as to enhance coverage efficiency. The transition towards fully covered and accessible environments may become disruptive. Hence, real-life applications may be facilitated, and other parameters such as medium blockage can be taken into account.

2.2.3.2 Network Connectivity

In the context of Drone deployment and in comparison with area coverage objective that—in some literature—is designed according to the specific application requirements, network connectivity is an IoT-inherent terminology that greatly depends on interoperability between all networking paradigms. In Ref. [46], authors propose an IoT-based UAV planning platform that ensures integration of SC domains such as smart grid (SG) and smart transportation systems (STSs). To this extent, Drones can even be used similar to roadside units in vehicular ad hoc networks [47,48].

A well-planned network connectivity ensures a good range of network coverage. The integration of different networking paradigms is an essential consideration in this objective. SG, STSs, and other SC applications can be planned simultaneously for improving the network performance.

2.2.3.3 Network Lifetime

Unlike stationary sensor nodes that can take decisions on when to turn on or off [49], Drones are battery-based wireless flying devices that are not connected to any infrastructure, which makes them more vulnerable to failures. Therefore, it is a big challenge to power the Drones for long durations of time. In this regard, Ref. [50] proposes a wireless Drone powering scheme in which the Drone stores harvested energy while flying to use it when hovering or in stationary mode. Authors in Ref. [51] target lifetime in mobile sensor networks as an optimization trade-off between lifetime and detection latency in surveillance applications. Furthermore, a weighted-clustering algorithm is employed in Ref. [52] to improve UAV-based ad hoc network parameters, such as number of cluster reduction, Drone lifetime, and network endurance. In the light of WSNs, authors in Ref. [53] present a Drone model that acts as a relay node for data collection from cluster-based WSNs, where the CH is responsible for forwarding data to the Drone. As an outcome, authors develop a trade-off between maximum number of hops and energy consumption of Drones and sensors, which in return helps extending network lifetime.

In addition to Drone energy consumption and battery limitations, future research may consider weather condition impacts on Drone failure. That is to say, design parameters may differ from one application to another, such as deployment of Drones in environmental-harsh conditions (extreme high/low temperatures) for data collection. Drone resiliency against wind can be investigated as well.

2.2.3.4 Data Fidelity

Owing to the vast proliferation of data-forwarding devices and end-user premises, data congestion and redundancy are becoming huge challenges to utilities and network providers. Thus, it is essential to ensure a filtering medium between ground-level sources and data centers. For instance, in legacy networking, stationary relay nodes can provide communication between sensor nodes and base stations to enhance data transmission reliability by merging and filtering the redundant data collected from sensor nodes as data packets travel through the network [54]. Base-station mounted Drones offer a great alternative in this regard, where they can travel to the area of interest and hover while performing data-relaying tasks. The key advantage behind employing Drones in such application is their mobility. In sudden instances where data traffic occurs or connection is lost due to a natural disaster or weather conditions, Drone can be sent to the specific areas and help in data relaying or maintaining the lost connection.

2.3 Dynamic Positioning of Drones

In contrary with Drone static deployment, Drone dynamic positioning implies that a Drone or a set of Drones deployed in the desired region can move from point A to point B and vice versa. The mobility of the Drone introduces further considerations (such as speed, relocation ability, and movement trigger) that should be taken into account for optimal deployment. For this reason, we investigate dynamic Drone repositioning schemes and relocation issues in the following subsections.

2.3.1 Drones Repositioning Schemes

Wireless communication systems for coverage and data collection, pre- and post-disaster situation assessment, search and rescue, security and warfare, and aerial imagery are some of the wide applications of Drones. Given the nature of these applications and the condition of the environment where they are deployed, it would require Drones to fly along a certain designated path [55–57]. Fast deployments and the ability to relocate Drones, either (1) schedule-based or (2) on demand, are the selling points of Drones compared with on-the-ground alternatives. Table 2.3 provides a comparison of dynamic Drone repositioning schemes.

Drones according to the deployment characteristics could act as competing and/or collaborative Drones for the nature of the objective of deployment, each having its own benefits and drawbacks. For example, Drones competing each other to increase their individual coverage [58] or when swarms of Drones collaborate together to increase mobility and reduce search time and search distance [59], or to increase robustness in the collected data [60,61], or even for collision avoidance and path optimization [62].

In addition to the earlier advantages, increase in network lifetime and communication quality [63] and the possibility of increase in node lifetime by introducing a charging station for recharging node from a single Drone [64] are other capabilities of Drones.

Many techniques and algorithms are used throughout the literature for efficient deployment of Drones, with the primary objective being increase coverage, throughput, and energy optimization as well as secondary objective of the minimum Drone requirements for a given application. A noncooperative game theory was used in Ref. [58] for deployments of Drones in temporary areas on a scheduled-based reposition. Similarly, authors in Ref. [65] introduced DroneCells; a flying base station for serving mobile users on the ground with on-demand relocation for improving 5G spectral efficiency. In another study by the same authors [66], it was shown that by freeing Drones in the cell of the network grids by letting them to move freely, larger system throughput can be achieved. However, user association scheme is a cost of the performance gain of their model, which required further investigation. Drones positing in 5G network was discussed in Ref. [67]. The authors utilized matrix-coloring approach for user equipment (UE) to UAV mapping by applying self-healing neural model for high-throughput coverage. In another study [68], for maximizing throughput, a mobile relaying technique was proposed. Compared with the conventional static relaying methods, mobile relays can be more cost effective while maintaining high mobility as well as easier adjustment of the nodes to offer a delay-tolerant network. In their study by optimizing relay trajectory and source/relay power allocation through an iterative algorithm, they achieved maximized throughput. Another use of an iterative algorithm for Drone placement was studied in Ref. [69]. Their approach considers UAVs as remote antenna units (RAUs) or remote radio heads (RRHs) serving as relay for transferring messages from mobile users to a central processor in the scheme of coordinate multipoint (CoMP). UAVs as a mobile cloudlet for enabling intensive computations on the cloud and as mobile base station for providing wireless services to ground was proposed in Refs. [70,71], respectively. These two approaches using optimal bit allocation and optimal transport theory achieved maximum efficiency while offering a significant energy saving. In Ref. [63], authors proposed a mobile data collector from wake-up schedule-based sensor nodes and optimizing UAV's trajectory

Table 2.3 A Comparison of Dynamic Drone Repositioning Schemes

Scheme	Ref.	Application	Algorithm	Primary Objective	Secondary Objective
Schedule-based	[58]	UAV activity scheduling	Game theory	Energy consumption optimization	-
	[63]	Mobile data collector for SN	Successive convex iteration	Network energy saving	-
	[73]	Environment-aware DBSs	Building detection algorithms	Coverage and energy efficiency	-
	[76]	Impact analysis after a disaster	-	Produce georeferenced data	-
	[78]	Uplink data collection from ground IoT devices	Novel framework	Minimum total transmit power	-
	[79]	Data collection from CHs	Resource allocation mechanism	Minimize CHs transmit power	Minimum number of UAV
	[80]	UAV-aided wireless communications	Cyclical multiple access	Max-Min throughput	-
On demand	[65]	DBSs	Game theory	5G spectral efficiency	-
	[67]	5G UE to UAV mapping	Self-healing neural model	Throughput coverage	-
	[68]	Mobile relaying	Iterative algorithm	Maximized throughput	-
	[69]	Flying RAUs or RRHs as relays	An iterative algorithm	Maximum average throughput	-

(Continued)

Table 2.3 (*Continued*) A Comparison of Dynamic Drone Repositioning Schemes

Scheme	Ref.	Application	Algorithm	Primary Objective	Secondary Objective
	[70]	Moving cloudlet	Optimal bit allocation	Significant energy savings	-
	[71]	Deployment framework of mobile base stations	Optimal transport theory	Minimum power consumption	-
	[72]	Providing access to core/backbone network	MAD-P	Minimal centralized control	Serving most data traffic demand-throughput
	[74]	Aerial LTE base station	REMs	Maximize coverage	-
	[75]	Power network damage assessment	Two-phase mathematical optimization model	Minimizing setup operating cost	-
	[77]	Ground-to-air uplink	Adaptive heading algorithm	Maximize SINR	-
	[81]	5G Backhaul-aware Drone placement	Backhaul limited	Maximum user served/maximum sum rate	Network robustness with respect to user movement
	[82]	Real-time object tracking (vehicles)	Various computation-intensive algorithms	Tracking objects with GPS exchange	-

(Continued)

Table 2.3 (*Continued*) A Comparison of Dynamic Drone Repositioning Schemes

Scheme	Ref.	Application	Algorithm	Primary Objective	Secondary Objective
	[25]	Cellular communication in a disaster-affected area	Polygon area decomposition	Maximum coverage of users	-
	[83]	Provide service to mobile users in a CRAN system	Novel algorithm	Optimizing QoE of uses	Minimum total transmit power
	[84]	Crowd-based Drone deployment	Coulomb's law	Backbone for network connectivity	-
	[66]	Higher performance in cellular network with DBS	Game theory	Larger system throughput	

for an overall network energy saving. An autonomous Drone capable of position and channel configuration with minimum centralized control was introduced in Ref. [72]. Their proposed measurement-aided dynamic planning (MAD-P) technique for self-configuration of Drones on the fly achieves maximum throughput and most data traffic served. Adaptive repositioning of Drones with changes to the terrain and environment while offering wireless and LTE coverage was studied in Refs. [73,74]. While in Ref. [73], Google Earth Engine with building detection algorithms was used for relocation of Drones [74], and radio environment maps (REMs) was used to construct a 3D space for maximum coverage. An interesting research in prepositioning of Drones for power network damage assessment with the anticipation of incoming extreme weather was proposed in Ref. [75]. An application for early disaster analysis using Drone was studied in Ref. [76] with the objective to gather georeferenced data about the impacted area. The Drones were controlled by Drone global positioning system (DGPS) and satellite-based augmentation systems. Another application in a disaster-affected area for deployment of base stations was studied in Ref. [25]. The author utilized polygon area decomposition technique as well as collision avoidance with other Drones to provide maximum user coverage. An adaptive heading algorithm was used in Ref. [77] along with beamforming for relay positioning of UAV for maximizing signal-to-interference-plus-noise ratio (SINR) resulting in optimizing the uplink communication performance. The other approach of deployment and reposition of Drones are based on minimizing the total transmit uplink power from IoT devices or CH. In Ref. [78] authors presented a novel framework where the movement of Drones and its 3D trajectory are based on the time-varying IoT network and the device activation process. Similarly in Ref. [79], an optimal scheduling and resource allocation technique was used for a power-efficient flying base station in CH-based machine to machine (M2M) communications. In Ref. [80], authors presented a cyclical wireless communication schedule between Drones and ground terminals. Their results showed a significant advantage in maximizing throughput over static UAV mobile base stations in wireless ad hoc applications. Authors in Ref. [81] used backhaul-limited 3D optimal Drone placement algorithm to increase coverage of 5G network in an area. Two different backhaul network, namely network-centric and user-centric were tested, while in the former user maximization and in the latter sum-rate maximization were achieved. A real-time object tracking via Drones was proposed in Ref. [82]. Authors in their study proposed a global positioning system (GPS) location-aware Drone for tracking moving object over the internet with great accuracy for slow-moving objects. By utilizing a cloud radio access network (CRAN) system, authors in Ref. [83] presented a novel framework to provide series to mobile users with an optimized quality of experience (QoE) with minimum power consumption. By incorporating machine-learning framework, they were able to predict user's content request distribution and their mobility pattern and cache the mentioned information to UAVs. A crowd-based UAV deployment for providing backbone for network connectivity was introduced in Ref. [84]. Authors believed that location-dependent deployments in case of moving users are inefficient. In order to prove their model effectiveness, they compared their solution with k-means clustering technique showing a good performance of their model.

2.3.2 Relocation Issues

Any efficient deployment of Drones at bare minimum requires having a maximum coverage of ground devices, i.e. Drones acting as relays, in addition to providing maximum throughput while maintaining an overall efficient network power consumption. Moreover, adequate mobility with minimum power usage is another important criterion, which in many cases, is the trade-off between high coverage and lower update time for repositioning of Drones. An efficient model should be able to consider all these criteria. Extra conditions, such as optimal trajectory and collision avoidance, can also be implemented for a system that is more effective.

Issues such as reporting wrong or falsified location of users to the Drones [85], limited battery lifetime of Drones while relocating from one position to another, and the possibility of collision in civilian airspace have to be kept in mind while implementing a UAV-involved application. Feasibility in multi-Drone deployment and environmental conditions affecting the performance of Drones [75] are other factors that add to the complexity of Drone deployments and repositioning, including sophisticated algorithms as well as overhead computations infeasibility.

2.4 Performance Metrics in Deployments

It is essential to evaluate the performance of the optimized deployment strategy. Cost, connectivity, and energy are three significant parameters. For instance, Ref. [86] targets cost and connectivity of Drone deployment for vessel emissions monitoring. Ref. [87] also targets and connectivity for optimizing the spatial location of medical Drones. Furthermore, Ref. [88] investigates multiple Drone deployment in radio access networks considering cost and connectivity of deployment. In contrary, Ref. [89] targets cost and energy parameters for target coverage applications, as shown in Table 2.4.

To this extent, future research proposals may consider the different performance metrics in the network to enhance deployment efficiency.

2.5 Existing Drone Brands and Deployment Issues

With the significant proliferation in Drone demand, many Drone brands emerged. In this work, we present the most commonly used ones. Table 2.5 compares those brands according to their different design characteristics (brands' data retrieved

Table 2.4   Performance Metrics and Deployments Comparison

	Considered Performance Metrics			Deployment Approach	
Ref.	Cost	Connectivity	Energy	Purpose	Targeted Place
[86]	✓	✓	-	Monitoring	Vessels emissions
[87]	✓	✓	-	Deployment location	Medicine
[89]	✓	-	✓	Target coverage	No specific areas
[88]	✓	✓	-	Assisting radio access networks	Cities
	✓	✓	✓	Monitoring	ITS

Table 2.5   Drone Brands Comparison

Specification	DJI Phantom 3	Veho Muvi Q-1	AEE Technology AP11	DJI Mavic pro	X Star Premium	DJI Phantom 4 pro
Number of rotors	4	4	4	4	4	4
Max. takeoff weight (g)	1,236	1,650	1,650	743	1,600	1,375
Max. horizontal speed (m/s)	16	20	20	18	16	20
Max. ascent speed (m/s)	5	6	6	5	6	6
Max. descent speed (m/s)	3	4	4	3	3	3
Endurance (min)	25	20	20	27	25	30
Max. range (km)	24	24	24	29	24	36
Max. service ceiling (m)	6,000	-	-	5,000	1,000	6,000
Cost ($)	513	425	699	999	999	1,499

from Refs. [90–94]). Veho Muvi Q-1 and agent execution renvironment (AEE) technology AP11 have the highest takeoff weight among the brands, whereas DJI Mavic pro is the lightest. In terms of speed, Veho Muvi Q-1 and AEE technology AP11 are the fastest; DJI Phantom 3 is relatively the slowest, and the remaining technology fall in between. However, DJI Phantom 4 pro is characterized with the highest endurance time, maximum range that it can reach, and service ceiling, which undoubtedly increases the brand cost. Despite the various advantages that these brands have, they need to be enhanced in terms of operation in windy weather conditions. Their battery capacity ranges from 3,830 to 6,800 mAh, and hence their endurance time is limited accordingly. Owing to the application-different requirements (more than 30 min of endurance), energy storage of these devices may need further attention by utilizing new techniques to enhance their endurance. To this extent, solar energy harvesting units can be included in the design of Drones as well as backup battery units to sustain power supply to the Drone. Furthermore, future Drones may be designed in such a way that enables mounting base stations for data relaying in communication networks as discussed earlier. In addition, several different types of UAVs are listed as follows [95]:

Target and decoy—providing ground and aerial gunnery, a target that is designed to mimic the radar and heat signature of an actual aircraft as to confuse or mislead enemy forces and weaponry.

Reconnaissance—used to gather key intelligence data on movements of potential threats, noncombatants, and friendly forces; help to ensure mission success during maneuvers at brigade level; and help in determining locations of threats and providing detailed topographic information in real time.

Combat—used in many forms of aerial attacks with varying capabilities, such as stealth attacks in which the Drone can sneak undiscovered and deal damage to enemy forces or can be used to just deploy bombs or missiles directly as well as other forms of weaponry.

R&D—used in research and development to further develop UAV technologies to be integrated into field-deployed aircrafts.

Civilization—UAVs are so versatile when it comes to their usages, some of them are used in 3D mapping, in which Drones take many pictures that help them create an accurate representation of the terrain. Another usage is search and rescue; UAVs can be used to find stranded humans or animals and help locate them so that they may be rescued and saved.

2.6 Drone-Based Applications

Drones in SCs have been studied and deployed in many areas such as Intelligent Transportation System (ITS) [96], civil control security [97,98], and crowd management [99,100]. In each scenario, Drone repositioning may be required to achieve the objective. For example, in the case of ITS, flying Drones can act as mobile speed cameras, or track and record license plates, or even as mobile police chasers to purist suspects on the road. In case of public safety and security, Drones can be augmented with facial recognition capabilities and can move around in big places such as stadiums or concerts for surveillance or remote monitoring purposes. Crowd control and management in cases such as a disaster can be chaos, given the significance and the location of the disaster. Flying Drones in this case can be used as an early guidance system for people to move to a safe location until the response team arrives. It may even be required for UAV-SC applications, as mentioned previously, to collaborate where multiple groups of Drones will be needed for the execution of their missions to benefit the overall energy efficiency as well as maximum coverage and mobility. It is impossible without a decisive algorithm that can cover all the aspects of an efficient deployment to deploy a group of Drones. Therefore, one should apply all the measures to ensure a good dynamic harmony and coordination for a fully autonomous operation of Drones.

Drones are used and applied widely in many areas due to their many benefits; their applications can be classified as environmental and industrial. Each can be divided into small applications depending on the area applied on. Each area will be described stating its challenges, the limitations that it could face, and the advantages and the profits that can be gained from it. Furthermore, the Drone and user type will also be specified.

2.6.1 *Environmental Applications*

Table 2.6 shows the challenges, advantages, profits, and user types for environmental applications. We find similarities between irrigated land mapping, impervious surface mapping, and watershed planning. They share the same challenge in collecting all the required data and the amount of time involved in that process. The advantages are namely decreasing the crew utility cost while providing data on water systems according to the required scales. The profit of using Drones in hydrological services is the direct effectiveness and efficient methods that lead to cost reduction of users. Table 2.2 also shows a similarity, moreover, between biomass, forest health, and disease detection. These share the same challenges in forest dynamics, species detection, and forest disturbance evaluation. Drone technology can be used to detect pest and disease outbreaks in an early stage, which is a significant advantage. The main disadvantage of Drones is that they only last up to 25–30 min in the air. The main aim of forestry services and departments is sustainable management of forests. A similar result is observed in vegetation management, North American Electric Reliability Corporation (NERC) right of way (ROW) monitoring, and asset verification. These share the same challenge, which is sending out crews to carry out assessments and inspections regularly, and during disastrous weather conditions. The advantage of Drones in identifying clearance violations and hazardous trees, in addition to automated analysis, species

Table 2.6 Drone Environmental Applications

Application	Challenges	Advantages	Limitations	Profit	User Type	Drone Type
Wildfire	Unauthorized Drones threat [101]	Safer [102]	Governmental flight control [103]	Increased safety	Public emergency management services	Civil and commercial Drones
Flooding	Needs to be coordinated [104]	Affordable and faster response [104]	Cannot assess damage clearly	Increase safety	Public emergency management services.	Civil and commercial Drones
Damage assessment	Qualitative data	Mapping damaged areas	Complex technology	Cost reduction	Public emergency management services	Civil and commercial Drones
Rapid response	Technology	Accurate live data	Number of Drones	Cost reduction	Public emergency management services.	Reconnaissance
Vegetation management	Carrying out assessments [105]	Vegetation assessment [106]	Covering huge land areas	Cost reduction	Electrical companies	Civil and commercial Drones
NERC ROW monitoring	Carrying out assessments [105]	Reduction in cost [105]	Covering huge land areas	Cost reduction	Land developers	Civil and commercial Drones
Asset verification	Carrying out assessments [105]	Reduce human risk [106]	Covering huge land areas	Cost reduction	Forest management	Civil and commercial Drones

(Continued)

Table 2.6 (Continued) Drone Environmental Applications

Application	Challenges	Advantages	Limitations	Profit	User Type	Drone Type
Biomass	Forest dynamics [107]	High accuracy [107]	Battery life [108]	Sustainable management	Forestry service	Reconnaissance
Forest health	Forest dynamics [107]	Flexible data [107]	Battery life [108]	Sustainable Management	Forestry service	Reconnaissance
Disaster detection	Forest dynamics [107]	Disease detection [107]	Battery life [108]	Sustainable management	Forestry service	Reconnaissance
Land cover mapping	Difficulty of comparison of maps [109]	Greater accuracy [110]	Good land cover is extremely expensive [111]	Better spatial imagery	Mining and energy	Civil uses and
Carbon capping	Better integration of satellite usage [112]	Continuous spatial data [113]	Battery life	Cost reduction	Mining and energy	Commercial
Renewable energy	Renewable energy firms are risky [114]	Increase in efficiency [115]	Battery life.	Increase safety	Energy companies	Research and

Table 2.6 (Continued) Drone Environmental Applications

Application	Challenges	Advantages	Limitations	Profit	User Type	Drone Type
Irrigated land mapping	Collecting data [113]	Continuous spatial data [113]	Water management [113]	Effective	Hydrological services	development
Impervious surface mapping	Collecting data [113]	Continuous spatial data [113]	Water management [113]	Efficient method	Hydrological services	Research and development
Watershed planning	Collecting data [113]	Continuous spatial data [113]	Water management [113]	Cost reduction	Hydrological services	development
Crop type	Plant count [116]	Faster and more efficient [117]	Battery life span	Saves time	Insurance agencies, farmers	Civil and commercial UAVs
Plant count	Plant count [116]	Faster and more efficient [117]	Battery life span	Saves time	Insurance agencies, farmers	Civil and Commercial
Leaf area index	Sensor-based methods	Faster and more efficient [117]	Battery life span	Reduces cost	Insurance agencies, farmers	UAVs
Growth stage	Sensor-based methods	Faster and more efficient [117]	Battery life span	Saves time	Insurance agencies, farmers	Reconnaissance
Plant height	Sensor-based methods	Faster and more efficient [117]	Battery life span	Reduces cost	Insurance agencies, farmers	Research and development

(Continued)

Table 2.6 (Continued) Drone Environmental Applications

Application	Challenges	Advantages	Limitations	Profit	User Type	Drone Type
Yield monitoring	Sensor-based methods	Faster and more efficient [117]	Battery life span	Saves time	Insurance agencies, farmers	Research and development
Plant health	Sensor-based methods	Faster and more efficient [117]	Battery life span	Saves time	Insurance agencies, farmers	Research and development
Nitrogen deficiencies	Determining the symptoms	Faster and more efficient [117]	Battery life span	Saves time	Insurance agencies, farmers	Research and development
Soil type	Identifying soil types.	Faster and more efficient [117]	Battery life span	Decrease cost	Insurance agencies, farmers	Research and development
Soil moisture	Identifying soil moisture	Faster and more efficient [117]	Battery life span	Decrease cost	Insurance agencies, Farmers	Research and development

recognition, growth forecasting, and normalized difference vegetation index give Drones the upper hand in comparison to human resources. Electrical companies, land developers, forest management, and NERC need more than one Drone to carry out inspections along the legal guidelines on how high a UAV can fly; this limits the field of perspective of the Drone.

Another advantage is that when Drones scan territories, land cover mapping and carbon capping have greater accuracy when scanning the changing environment and landscape in comparison to other alternatives. Drones are also more efficient for solar panel installation. The main advantage is having better spatial imagery, cost reduction, better safety, and security for the crew. Furthermore, the table shows that different applications in disaster responses have different challenges, advantages, profits, limitations, and user types. The advantages are that Drones are safer than helicopters in thick smoke areas and more affordable. In addition, in monitoring and mapping damaged areas within the vicinity of disaster areas, Drones are more effective. The drawbacks of Drones in disaster response is mainly legal, where the government views Drones as potential risks, thus preventing Drones from assessing damage completely while complying with the regulations that limit the number of Drones to be operated. Under environmental, we can observe that there are ten applications for agriculture that gives a wide range of data results. For example, we can see that there is a similarity in the challenges, generally, such that it is hard to collect data and define the problems in large areas. The advantage, however, is that Drones minimize the time of data collection and in an efficient manner. The main limitation of the Drones is battery capacity.

2.6.2 Industrial Applications

Table 2.7 shows that Drone deployment in different applications of oil and gas industry face different challenges, have different advantages, profits, limits, and user types as well. The main advantage of using Drones in these fields is that Drones can effectively detect and track the area of pollution and investigate air emission to obtain required data. Using Drones is efficient, cost effective, safer, and secure.

Putting the previous subsections into consideration, Figure 2.3 categorizes Drone deployment segments including type, application area, and objective.

2.7 Open Research Issues

According to the analysis of literature, we present the following open research gaps that future efforts may consider:

■ Facilitation of random and controlled deployment methodologies so as to enhance network performance.

Table 2.7   Drone Industrial Applications

Application	Challenges	Advantages	Limitations	Profit	User Type	Drone Type
Oil spill tracking	Maritime activities [118]	Detecting, tracking and modeling the area of pollutants [118].	Only limited to monitoring.	Efficiency	Petroleum companies	Reconnaissance
Environmental assessment	Monitoring ecosystem health	Air emission Investigations [119]	Safety and privacy legal restraints	Cost reduction	Petroleum companies	Reconnaissance
Pipeline monitoring	Leak detection [120]	Thermal diagnostics [121] and conduct surveying [122]	Drones can only identify the problem	Production stability	Energy companies	Reconnaissance

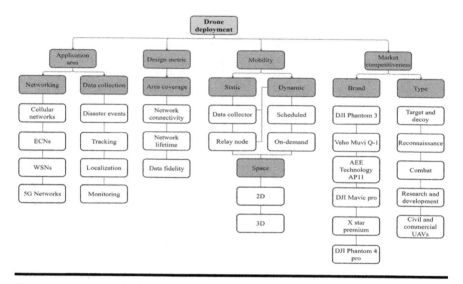

Figure 2.3   Drone deployment taxonomy.

■ Development of integrated optimization approaches that consider the trade-offs between the aforementioned application objectives such as maximum coverage at the cost of power consumption or number of Drones.
■ Design of scalable Drone-based networks since the number of users and ground nodes are exponentially increasing.
■ Assessment of medium blockage effects on Drone range of coverage.
■ Simultaneous integration of different networking paradigms, such as smart transportations (STs), SGs, and SCs in order to achieve better network connectivity.
■ Assessment of wind and weather condition effects on Drone robustness against harsh environmental conditions to enhance network lifetime.
■ Design of data-intensive and time-sensitive Drones, especially for ECNs, where Drone efficiency is crucial.
■ Optimization of dynamic Drone trajectory in 3D spaces for better accuracy and speed.
■ Facilitation of automatic collision avoidance capabilities in dynamic Drones.
■ Enforcement of Drone-embedded data filtering techniques to avoid reporting false locations and limit duplicate or redundant data.
■ Drone battery lifetime enhancement, especially while relocating. In this regard, energy harvesting techniques and solar energy storage can be employed to store energy in a backup battery.

2.8 Conclusion

Recently, Drones gained an increased attention due to their relocation capabilities and use flexibility. However, their deployment strategies are diverse, depending on the application requirements. In this work, we provide an in-depth understanding of Drone deployment types, including static and dynamic deployments, by revealing the related ongoing research efforts. We further shed light on existing Drone brands and their deployment issues. Then, we investigate literature and propose open research issues accordingly.

References

1. F. Al-Turjman, E. Ever, and H. Zahmatkesh, Small cells in the forthcoming 5G/IoT: Traffic modelling and deployment overview, *IEEE Communications Surveys and Tutorials*, 2018. doi:10.1109/COMST.2018.2864779.
2. M. Gharibi, R. Boutaba, and S. L. Waslander, Internet of Drones, *IEEE Access*, 2016. doi:10.1109/ACCESS.2016.2537208.
3. S. Chandrasekharan et al., Designing and implementing future aerial communication networks, *IEEE Communications Magazine*, 2016. doi:10.1109/MCOM.2016.7470932.
4. F. Al-Turjman, and S. Alturjman, 5G/IoT-enabled UAVs for multimedia delivery in industry-oriented applications, *Springer's Multimedia Tools and Applications Journal*, 2018. doi:10.1007/s11042-018-6288-7.
5. E. Casella et al., Study of wave runup using numerical models and low-altitude aerial photogrammetry: A tool for coastal management, *Estuarine, Coastal and Shelf Science*, 2014. doi:10.1016/j.ecss.2014.08.012.
6. J. Lisein, M. Pierrot-Deseilligny, S. Bonnet, and P. Lejeune, A photogrammetric workflow for the creation of a forest canopy height model from small unmanned aerial system imagery, *Forests*, 2013. doi:10.3390/f4040922.
7. R. I. Bor-Yaliniz, A. El-Keyi, and H. Yanikomeroglu, Efficient 3-D placement of an aerial base station in next generation cellular networks, in *2016 IEEE International Conference on Communications, ICC*, Kuala Lumpur, Malaysia, 2016.
8. M. Mozaffari, W. Saad, M. Bennis, and M. Debbah, Drone small cells in the clouds: Design, deployment and performance analysis, in *2015 IEEE Global Communications Conference, GLOBECOM*, San Diego, CA, 2015.
9. A. Mazayev, N. Correia, and G. Schütz, Data gathering in wireless sensor networks using unmanned aerial vehicles, *International Journal of Wireless Information Networks*, vol. 23, no. 4, 297–309, December 2016.
10. D. Floreano, and R. J. Wood, Science, technology and the future of small autonomous Drones, *Nature*, vol. 521, no. 7553, 460–466, May 2015.
11. K. Nonami, Drone technology, cutting-edge Drone business, and future prospects, *Journal of Robotics and Mechatronics*, 2016. doi:10.20965/jrm.2016.p0262.
12. N. Hossein Motlagh, T. Taleb, and O. Arouk, Low-altitude unmanned aerial vehicles-based internet of things services: Comprehensive survey and future perspectives, *IEEE Internet of Things Journal*, 2016. doi:10.1109/JIOT.2016.2612119.

13. L. Tang and G. Shao, Drone remote sensing for forestry research and practices, *Journal of Forestry Research*, 2015. doi:10.1007/s11676-015-0088-y.
14. H. Yang, Y. Lee, S. Y. Jeon, and D. Lee, Multi-rotor Drone tutorial: systems, mechanics, control and state estimation, *Intelligent Service Robotics*, 2017. doi:10.1007/s11370-017-0224-y.
15. M. Hassanalian and A. Abdelkefi, Classifications, applications, and design challenges of Drones: A review, *Progress in Aerospace Sciences*, 2017. doi:10.1016/j.paerosci.2017.04.003.
16. M. Hassanalian, D. Rice, and A. Abdelkefi, Evolution of space Drones for planetary exploration: A review, *Progress in Aerospace Sciences*, 2018. doi:10.1016/j.paerosci.2018.01.003.
17. A. Otto, N. Agatz, J. Campbell, B. Golden, and E. Pesch, Optimization approaches for civil applications of unmanned aerial vehicles (UAVs) or aerial Drones: A survey, *Networks*, 2018. doi:10.1002/net.21818.
18. J. S. Patel, F. Fioranelli, and D. Anderson, Review of radar classification and RCS characterisation techniques for small UAVs or Drones, *IET Radar, Sonar and Navigation*, 2018. doi:10.1049/iet-rsn.2018.0020.
19. T. Rakha and A. Gorodetsky, Review of Unmanned Aerial System (UAS) applications in the built environment: Towards automated building inspection procedures using Drones, *Automation in Construction*, 2018. doi:10.1016/j.autcon.2018.05.002.
20. K. Bhatt, A. Pourmand, and N. Sikka, Targeted applications of unmanned aerial vehicles (Drones) in telemedicine, *Telemedicine Journal and e-Health*, vol. 24, no. 11, 833–838, November 2018.
21. E. Tuba, I. Tuba, D. Dolicanin-Djekic, A. Alihodzic, and M. Tuba, Efficient Drone placement for wireless sensor networks coverage by bare bones fireworks algorithm, in *6th International Symposium on Digital Forensic and Security, ISDFS 2018- Proceeding*, Antalya, Turkey, 2018.
22. I. Strumberger, N. Bacanin, S. Tomic, M. Beko, and M. Tuba, Static Drone placement by elephant herding optimization algorithm, in *2017 25th Telecommunications Forum, TELFOR 2017- Proceedings*, Belgrade, Serbia, 2018.
23. O. Menéndez, M. Pérez, and F. Auat Cheein, Visual-based positioning of aerial maintenance platforms on overhead transmission lines, *Applied Sciences*, vol. 9, no. 1, 165, January 2019.
24. P. V. Klaine, J. P. B. Nadas, R. D. Souza, and M. A. Imran, Distributed Drone base station positioning for emergency cellular networks using reinforcement learning, *Cognitive Computation*, 2018. doi:10.1007/s12559-018-9559-8.
25. X. Li, Deployment of Drone base stations for cellular communication without apriori user distribution information, in *Chinese Control Conference, CCC*, Wuhan, China, 2018, pp. 7274–7281.
26. X. Li and L. Xing, Optimal deployment of Drone base stations for cellular communication by network-based localization, in *Chinese Control Conference, CCC*, Wuhan, China, 2018, pp. 7282–7287.
27. F. Lagum, I. Bor-Yaliniz, and H. Yanikomeroglu, Strategic densification with UAV-BSS in cellular networks, *IEEE Wireless Communications Letters*, vol. 7, no. 3, 384–387, June 2018.
28. M. Deruyck, J. Wyckmans, W. Joseph, and L. Martens, Designing UAV-aided emergency networks for large-scale disaster scenarios, *EURASIP Journal on Wireless Communications and Networking*, 2018. doi:10.1186/s13638-018-1091-8.

29. Report ITU-R M.2135-1, (ITU-R M.2135-1) Guidelines for evaluation of radio interface technologies for IMT advanced, Evaluation, 2009.
30. I. Bor-Yaliniz, S. S. Szyszkowicz, and H. Yanikomeroglu, Environment-aware Drone-base-station placements in modern metropolitans, *IEEE Wireless Communications Letters*, 2018. doi:10.1109/LWC.2017.2778242.
31. C. Dong, J. Xie, H. Dai, Q. Wu, Z. Qin, and Z. Feng, Optimal deployment density for maximum coverage of Drone small cells, *China Communications*, vol. 15, no. 5, 25–40, 2018.
32. A. M. Hayajneh, S. A. R. Zaidi, D. C. McLernon, and M. Ghogho, Drone empowered small cellular disaster recovery networks for resilient smart cities, in *2016 IEEE International Conference on Sensing, Communication and Networking, SECON Workshops*, London, 2016, pp. 1–6.
33. M. Y. Selim, A. Alsharoa, and A. E. Kamal, Hybrid cell outage compensation in 5g networks: Sky-ground approach, in *IEEE International Conference on Communications*, Kansas City, MO, 2018, pp. 1–6.
34. M. Gapeyenko, I. Bor-Yaliniz, S. Andreev, H. Yanikomeroglu, and Y. Koucheryavy, Effects of blockage in deploying mmWave Drone base stations for 5g networks and beyond, in *2018 IEEE International Conference on Communications Workshops, ICC Workshops- Proceedings*, Kansas City, MO, 2018, pp. 1–6.
35. A. Merwaday and I. Guvenc, UAV assisted heterogeneous networks for public safety communications, in *2015 IEEE Wireless Communications and Networking Conference Workshops, WCNCW*, New Orleans, LA, 2015, pp. 329–334.
36. A. Akarsu and T. Girici, Fairness aware multiple Drone base station deployment, *IET Communications*, vol. 12, no. 4, 425–431, March 2018.
37. L. Wang, B. Hu, and S. Chen, Energy efficient placement of a Drone base station for minimum required transmit power, *IEEE Wireless Communications Letters*, pp. 1–1, 2018.
38. B. S. Morse, C. H. Engh, and M. A. Goodrich, UAV video coverage quality maps and prioritized indexing for wilderness search and rescue, in *Proceeding of the 5th ACM/IEEE international conference on Human-robot interaction - HRI '10*, Osaka, Japan, 2010, pp. 227–234.
39. J. M. Boehmler et al., Development of a multispectral albedometer and deployment on an unmanned aircraft for evaluating satellite retrieved surface reflectance over Nevada's black rock desert, *Sensors (Switzerland)*, vol. 18, no. 10, 3504, October 2018.
40. F. Al-Turjman, and S. Alturjman, Context-sensitive access in Industrial Internet of Things (IIoT) healthcare applications, *IEEE Transactions on Industrial Informatics*, vol. 14, no. 6, 2736–2744, 2018.
41. R. Jackisch, S. Lorenz, R. Zimmermann, R. Möckel, and R. Gloaguen, Drone-borne hyperspectral monitoring of acid mine drainage: An example from the Sokolov lignite district, *Remote Sensing*, 2018. doi:10.3390/rs10030385.
42. Q. Lin, H. Song, X. Wang, and Z. Ouyang, Collaborative unmanned aerial systems for effective and efficient airborne surveillance, in *Disruptive Technologies in Information Sciences*, Orlando, FL, 2018, p. 14.
43. C. Kyrkou, G. Plastiras, T. Theocharides, S. I. Venieris, and C. S. Bouganis, Dronet: Efficient convolutional neural network detector for real-time UAV applications, in *Proceedings of the 2018 Design, Automation and Test in Europe Conference and Exhibition, DATE*, Dresden, Germany, 2018, pp. 967–972.

44. F. Betti Sorbelli, C. M. Pinotti, and V. Ravelomanana, Range-free localization algorithm using a customary Drone, in *Proceedings -2018 IEEE International Conference on Smart Computing, SMARTCOMP*, Taormina, Italy, 2018, pp. 9–16.

45. A. Al-Hourani, S. Kandeepan, and S. Lardner, Optimal LAP altitude for maximum coverage, *IEEE WirelesCommunications Letters*, vol. 3, no. 6, 569–572, December 2014.

46. A. Giyenko, and Y. I. Cho, Intelligent UAV in smart cities using IoT, in *International Conference on Control, Automation and Systems*, Gyeongju, South Korea, 2016, pp. 207–210.

47. M. Ben Brahim, W. Drira, and F. Filali, Roadside units placement within city-scaled area in vehicular ad-hoc networks, in *2014 International Conference on Connected Vehicles and Expo, ICCVE 2014- Proceedings*, Vienna, Austria, 2014, pp. 1010–1016.

48. Y. Liang, H. Liu, and D. Rajan, Optimal placement and configuration of roadside units in vehicular networks, in *IEEE Vehicular Technology Conference*, Yokohama, Japan, 2012, pp. 1–6.

49. M. N. Islam, Y. M. Jang, S. Choi, S. Park, and H. Park, Redundancy reduction protocol with sensing coverage assurance in distributed wireless sensor networks, in *2009 9th International Symposium on Communications and Information Technology*, Incheon, South Korea, 2009, pp. 631–636.

50. J. A. L. Calvo, G. Alirezaei, and R. Mathar, Wireless powering of Drone-based MANETs for disaster zones, in *2017 IEEE International Conference on Wireless for Space and Extreme Environments (WiSEE)*, Montreal, QC, 2017, pp. 98–103.

51. C. Wang, P. Ramanathan, and K. K. Saluja, Modeling latency—Lifetime trade-off for target detection in mobile sensor networks, *ACM Transactions on Sensor Networks*, 2010. doi:10.1145/1806895.1806903.

52. X. Xin et al., A weighted clustering algorithm based on node energy for multi-UAV Ad Hoc networks, in *Tenth International Conference on Information Optics and Photonics*, Beijing, China, 2018, p. 172.

53. C. Y. Tazibt, M. Bekhti, T. Djamah, N. Achir, and K. Boussetta, Wireless sensor network clustering for UAV-based data gathering, in *2017 Wireless Days*, Porto, Portugal, 2017, pp. 245–247.

54. U. Roedig, A. Barroso, and C. J. Sreenan, Determination of aggregation points in wireless sensor networks, in *Proceedings. 30th Euromicro Conference,* Rennes, France, 2004, pp. 503–510.

55. F. Belkhouche, Reactive optimal UAV motion planning in a dynamic world, *Robotics and Autonomous Systems*, vol. 96, 114–123, October 2017.

56. J. Cui, R. Wei, Z. Liu, and K. Zhou, UAV motion strategies in uncertain dynamic environments: A path planning method based on Q-learning strategy, *Applied Sciences*, vol. 8, no. 11, 2169, November 2018.

57. J. J. Ruz, O. Arevalo, G. Pajares, and J. M. De La Cruz, Decision making among alternative routes for UAVs in dynamic environments, in *IEEE International Conference on Emerging Technologies and Factory Automation, ETFA*, Patras, Greece, 2007, pp. 997–1004.

58. S. Koulali, E. Sabir, T. Taleb, and M. Azizi, A green strategic activity scheduling for UAV networks: A sub-modular game perspective, *IEEE Communications Magazine*, vol. 54, no. 5, 58–64, May 2016.

59. I. K. Ha, A probabilistic target search algorithm based on hierarchical collaboration for improving rapidity of Drones, *Sensors (Switzerland)*, vol. 18, no. 8, 2535, August 2018.

60. C. Gomez and H. Purdie, UAV- based photogrammetry and geocomputing for hazards and disaster risk monitoring – A review, *Geoenvironmental Disasters*, vol. 3, no. 1, 23, December 2016.
61. P. A. Zientara, J. Choi, J. Sampson, and V. Narayanan, Drones as collaborative sensors for image recognition, in *2018 IEEE International Conference on Consumer Electronics, ICCE*, Las Vegas, NV, 2018, pp. 1–4.
62. J. Sun, J. Tang, and S. Lao, Collision avoidance for cooperative UAVs with optimized artificial potential field algorithm, *IEEE Access*, vol. 5, 18382–18390, 2017.
63. C. Zhan, Y. Zeng, and R. Zhang, Energy-efficient data collection in UAV enabled wireless sensor network, *IEEE Wireless Communications Letters*, vol. 7, no. 3, 328–331, June 2018.
64. J. Johnson, E. Basha, and C. Detweiler, Charge selection algorithms for maximizing sensor network life with UAV-based limited wireless recharging, in *2013 IEEE Eighth International Conference on Intelligent Sensors, Sensor Networks and Information Processing*, Melbourne, VIC, 2013, pp. 159–164.
65. A. Fotouhi, M. Ding, and M. Hassan, DroneCells: Improving 5G spectral efficiency using Drone-mounted flying base stations, in *IEEE Globecom Workshops (GC Wkshps)*, Singapore, July 2017, pp. 1–6.
66. A. Fotouhi, M. Ding, and M. Hassan, Service on demand: Drone base stations cruising in the cellular network, in *2017 IEEE Globecom Workshops, GC Workshops 2017-Proceedings*, Singapore, July 2017.
67. V. Sharma, D. N. K. Jayakody, and K. Srinivasan, On the positioning likelihood of UAVs in 5G networks, *Physical Communication*, vol. 31, 1–9, December 2018.
68. Y. Zeng, R. Zhang, and T. J. Lim, Throughput maximization for UAV-enabled mobile relaying systems, *IEEE Transactions on Communications*, 2016. doi:10.1109/TCOMM.2016.2611512.
69. L. Liu, S. Zhang, and R. Zhang, CoMP in the sky: UAV placement and movement optimization for multi-user communications, arXiv:1802.10371, February 2018.
70. S. Jeong, O. Simeone, and J. Kang, Mobile cloud computing with a UAV-mounted cloudlet: Optimal bit allocation for communication and computation, *IET Communications*, vol. 11, no. 7, 969–974, May 2017.
71. M. Mozaffari, W. Saad, M. Bennis, and M. Debbah, Optimal transport theory for power-efficient deployment of unmanned aerial vehicles, in *2016 IEEE International Conference on Communications (ICC)*, Kuala Lumpur, Malaysia, 2016, pp. 1–6.
72. N. Lu, Y. Zhou, C. Shi, N. Cheng, L. Cai, and B. Li, Planning while flying: A measurement-aided dynamic planning of Drone small cells, *IEEE Internet of Things Journal*, 1, 2018. doi:10.1109/JIOT.2018.2873772.
73. A. French, M. Mozaffari, A. Eldosouky, and W. Saad, Environment-aware deployment of wireless Drones base stations with Google Earth simulator, Kyoto, Japan, May 2018.
74. A. Chakraborty, E. Chai, K. Sundaresan, A. Khojastepour, and S. Rangarajan, SkyRAN: a self-organizing LTE RAN in the sky, in *Proceedings of the 14th International Conference on emerging Networking EXperiments and Technologies - CoNEXT '18*, Greece, 2018, pp. 280–292.
75. G. J. Lim, S. Kim, J. Cho, Y. Gong, and A. Khodaei, Multi-UAV pre-positioning and routing for power network damage assessment, *IEEE Transactions on Smart Grid*, vol. 9, no. 4, 3643–3651, July 2018.

76. H. Bendea, P. Boccardo, S. Dequal, F. G. Tonolo, D. Marenchino, and M. Piras, Low cost UAV for post-disaster assessment, *Proceedings of the XXI Congress. International Society for Photogrammetry and Remote Sensing.* Beijing, July 2008.

77. F. Jiang and A. L. Swindlehurst, Dynamic UAV relay positioning for the ground-to-air uplink, in *2010 IEEE Globecom Workshops, GC'10*, Miami, FL, 2010, pp. 1766–1770.

78. M. Mozaffari, W. Saad, M. Bennis, and M. Debbah, Mobile unmanned aerial vehicles (UAVs) for energy-efficient Internet of Things communications, *IEEE Transactions on Wireless Communications*, vol. 16, no. 11, 7574–7589, November 2017.

79. M. N. Soorki, M. Mozaffari, W. Saad, M. H. Manshaei, and H. Saidi, Resource allocation for machine-to-machine communications with unmanned aerial vehicles, in *2016 IEEE Globecom Workshops, GC Workshops 2016- Proceedings*, Washington, DC, 2016, pp. 1–6.

80. J. Lyu, Y. Zeng, and R. Zhang, Cyclical multiple access in UAV-aided communications: A throughput-delay tradeoff, *IEEE Wireless Communicatios Letters*, 2016. doi:10.1109/LWC.2016.2604306.

81. E. Kalantari, M. Z. Shakir, H. Yanikomeroglu, and A. Yongacoglu, Backhaul-aware robust 3D Drone placement in 5G+ wireless networks, in *2017 IEEE International Conference on Communications Workshops, ICC Workshops*, Paris, France, 2017, pp. 109–114.

82. A. Koubaa, and B. Qureshi, DroneTrack: Cloud-based real-time object tracking using unmanned aerial vehicles over the internet, *IEEE Access*, 2018. doi:10.1109/ACCESS.2018.2811762.

83. M. Chen, M. Mozaffari, W. Saad, C. Yin, M. Debbah, and C. S. Hong, Caching in the sky: Proactive deployment of cache-enabled unmanned aerial vehicles for optimized quality-of-experience, *IEEE Journal on Selected Areas in Communications*, 2017. doi:10.1109/JSAC.2017.2680898.

84. D. Rautu, R. Dhaou, and E. Chaput, Crowd-based positioning of UAVs as access points, in *2018 15th IEEE Annual Consumer Communications & Networking Conference (CCNC)*, Las Vegas, NV, 2018, pp. 1–6.

85. X. Xu, L. Duan, and M. Li, UAV placement games for optimal wireless service provision, in *2018 16th International Symposium on Modeling and Optimization in Mobile, Ad Hoc, and Wireless Networks (WiOpt)*, Shanghai, China, 2018, pp. 1–8.

86. J. Xia, K. Wang, and S. Wang, Drone scheduling to monitor vessels in emission control areas, *Transportation Research Part B: Methodological*, vol. 119, 174–196, January 2019.

87. A. Pulver, and R. Wei, Optimizing the spatial location of medical Drones, *Applied Geography*, vol. 90, 9–16, January 2018.

88. W. Shi et al., Multiple Drone-cell deployment analyses and optimization in Drone assisted radio access networks, *IEEE Access*, vol. 6, 12518–12529, 2018.

89. D. Zorbas, L. D. P. Pugliese, T. Razafindralambo, and F. Guerriero, Optimal Drone placement and cost-efficient target coverage, *Journal of Network and Computer Applications*, vol. 75, 16–31, November 2016.

90. Mavic Pro Specs. [Online]. Available: www.dji.com/mavic/specs#specs. [Accessed: 25-Jan-2019].

91. Phantom 3 SE Specs. [Online]. Available: www.dji.com/phantom-3-se/info?lang=cn#specs. [Accessed: 25-Jan-2019].

92. Phantom 4 Pro V2.0 Specs. [Online]. Available: www.dji.com/phantom-4-pro-v2/info#specs. [Accessed: 25-Jan-2019].

93. Q1 Drone Specs. [Online]. Available: www.veho-muvi.com/muvi_product/q-Drone/. [Accessed: 25-Jan-2019].
94. X-Star Premium Specs. [Online]. Available: https://autelDrones.com/collections/x-star-accessories. [Accessed: 25-Jan-2019].
95. The UAV. [Online]. Available: www.theuav.com/. [Accessed: 31-Jan-2019].
96. H. Menouar, I. Guvenc, K. Akkaya, A. S. Uluagac, A. Kadri, and A. Tuncer, UAV-enabled intelligent transportation systems for the smart city: Applications and challenges, *IEEE Communications Magazine*, vol. 55, no. 3, 22–28, March 2017.
97. F. Al-Turjman, and S. Alturjman, Confidential smart-sensing framework in the IoT era, *The Springer Journal of Supercomputing*, vol. 74, no. 10, 5187–5198, 2018.
98. I. Maza, F. Caballero, J. Capitán, J. R. Martínez-de-Dios, and A. Ollero, Experimental results in multi-UAV coordination for disaster management and civil security applications, *Journal of Intelligent & Robotics Systems*, vol. 61, no. 1–4, 563–585, January 2011.
99. N. H. Motlagh, M. Bagaa, and T. Taleb, UAV-based IoT platform: A crowd surveillance use case, *IEEE Communications Magazine*, vol. 55, no. 2, 128–134, February 2017.
100. A. Al-Sheary, and A. Almagbile, Crowd monitoring system using unmanned aerial vehicle (UAV), *Journal of Civil Engineering and Architecture*, vol. 11, no. 11, 1014–1024, November 2017.
101. Department of Forestry and Fire Management, Drones and wildfire. [Online]. Available: https://dffm.az.gov/fire/information/Drones-and-wildfire. [Accessed: 31-Jan-2019].
102. UAS Insights, Drones & wildfires – Benefits and risks. [Online]. Available: www.uasinsights.com/2017/10/16/Drones-wildfires-benefits-and-risks/. [Accessed: 31-Jan-2019].
103. Federal Aviation Administration, Drones and wildfires don't mix-period.
104. Ambienal Risk Analysis, Aerial Drones to predict and assess flood damage. [Online]. Available: www.ambientalrisk.com/natural-environment-research-council/. [Accessed: 31-Jan-2019].
105. T&DWorld, Flying high to improve vegetation management. [Online]. Available: www.tdworld.com/overhead-transmission/flying-high-improve-vegetation-management. [Accessed: 31-Jan-2019].
106. Border States Supply Solutions, Drone vegetation management a game-changer for electric utilities. [Online]. Available: https://solutions.borderstates.com/Drone-vegetation-management-for-utilities/. [Accessed: 31-Jan-2019].
107. T. P. Banu, G. F. Borlea, and C. Banu, The use of Drones in forestry, *Journal of Environmental Science and Engineering B*, vol. 5, no. 11, 557–562, November 2016.
108. Recode, Wireless charging could keep Drones in the air for much longer. [Online]. Available: www.recode.net/2016/10/12/13257790/wireless-charging-Drones-air-longer-solar-power-batteries. [Accessed: 31-Jan-2019].
109. M. Herold, P. Mayaux, C. E. Woodcock, A. Baccini, and C. Schmullius, Some challenges in global land cover mapping: An assessment of agreement and accuracy in existing 1 km datasets, *Remote Sensing of Environment*, vol. 112, no. 5, 2538–2556, May 2008.
110. K. Iizuka, M. Itoh, S. Shiodera, T. Matsubara, M. Dohar, and K. Watanabe, Advantages of unmanned aerial vehicle (UAV) photogrammetry for landscape analysis compared with satellite data: A case study of postmining sites in Indonesia, *Cogent Geoscience*, vol. 4, no. 1, 1–15, July 2018.

111. Melodies Project, The challenges of mapping land cover. [Online]. Available: www.melodiesproject.eu/content/challenges-mapping-land-cover. [Accessed: 31-Jan-2019].

112. T. Kuemmerle et al., Challenges and opportunities in mapping land use intensity globally, *Current Opinion in Environmental Sustainability*, vol. 5, no. 5, 484–493, October 2013.

113. World Meteorological Organization, New challenges of water resources management: The future role of CHy. [Online]. Available: https://public.wmo.int/en/bulletin/new-challenges-water-resources-management-future-role-chy. [Accessed: 31-Jan-2019].

114. Union of Concerned Scientists, Barriers to renewable energy technologies. [Online]. Available: www.ucsusa.org/clean-energy/renewable-energy/barriers-to-renewable-energy#.XFREY1wzbct. [Accessed: 31-Jan-2019].

115. Green Tech. Media, Why Drones are 'Game-Changing' for renewable energy. [Online]. Available: www.greentechmedia.com/articles/read/why-Drones-are-game-changing-for-renewable-energy#gs.Ql2ldE4S. [Accessed: 31-Jan-2019].

116. Canada Centre for Remote Sensing, Fundamentals of remote sensing. [Online]. Available: http://sar.kangwon.ac.kr/etc/fundam/chapter5/chapter5_3_e.html. [Accessed: 31-Jan-2019].

117. I. Kalisperakis, C. Stentoumis, L. Grammatikopoulos, and K. Karantzalos, Leaf area index estimation in vineyards from UAV hyperspectral data, 2D image mosaics and 3D canopy surface models, *ISPRS - International Archives of the Photogrammetry, Remote Sensing and Spatial Information Sciences*, vol. XL-1/W4, 299–303, August 2015.

118. F. Aznar, M. Sempere, M. Pujol, R. Rizo, and M. J. Pujol, Modelling oil-spill detection with swarm Drones, *Abstract and Applied Analysis*, vol. 2014, 1–14, 2014.

119. L. Satterlee, Climate Drones: A new tool for oil and gas air emission monitoring, *Environmental Law Reporter*, vol. 46, no. 12, 11–69, 2016.

120. Huffington Post, Using Drones to monitor oil pipelines. [Online]. Available: www.huffingtonpost.com/entry/using-Drones-to-monitor-oil-pipelines_us_59390907e4b014ae8c69ddd4. [Accessed: 31-Jan-2019].

121. WORKAWELL Thermal Imaging Systems, Pipeline inspection with thermal diagnostics. [Online]. Available: www.Drone-thermal-camera.com/Drone-uav-thermography-inspection-pipeline/. [Accessed: 31-Jan-2019].

122. DroneBelow, Drone in pipeline inspection. [Online]. Available: https://Dronebelow.com/Drones-in-pipeline-inspection/. [Accessed: 31-Jan-2019].

Chapter 3

Optimal Placement for 5G Drone-BS Using SA and GA

Fadi Al-Turjman, Joel Poncha Lemayian,
Sinem Alturjman, and Leonardo Mostarda

Antalya Bilim University

3.1 Introduction

There is a high demand for provisioning high quality of services (QoS) due to recent gigantic growth in everything, especially in the telecommunication sector. The rapid population growth has brought a number of challenges in telecommunications, including coverage and data traffic capacity. One promising way to mitigate some of these challenges is the utilization of intelligent systems towards smart projects such as smart cities, smart building, smart vehicles, and smart grids. Internet of things (IoT) is the interconnection of these smart projects with sensing, actuation, and computing capabilities via the internet. It is used to provide better services and resource management for the general population. However, the vast amount of data generated and collected requires the use of a powerful communication paradigm to guarantee the QoS in all these services. 3G and 4G have a few QoS advantages, such as low deployment cost, simplicity in management, extensive coverage, and high security. However, they do not support low-cost machine-type communications with high efficiency. This is an important feature for the future telecommunication, and this is because 3G and 4G were designed mainly for optimized broadband communication [1]. On the other hand, 5G is specifically designed to provide QoS to its users, which means that it is capable of providing maximum bandwidth, reduce latency, error rate, and uptime. Additionally, 5G has increased data rate, reduced delay, as well as enhanced cellular coverage. In health care, for instance, these advantages came in handy in improving the system for millions of people. Chen et al. [2] designed a personalized emotion-aware health-care system using 5G, which focuses on emotional care, particularly for children, the mentally ill, and the elderly people. The proposed system uses various IoT devices to capture images and speech signals from a patient in an intelligent environment, such as a smart home. This data is fed into an emotion detection module that processes the speech and image signals separately and then merges the results to produce a final score of the emotion. The score is further analyzed to determine whether the patient requires attention, if so, medics are alerted immediately. Additionally, Chen et al. states that 99.87% of emotional detection response was read accurately from all the experiments that were conducted. Furthermore, the authors in Ref. [3] state that the ability of 5G to focus on heterogeneous access technology has opened a plethora of possibilities. 5G has the ability to create an interconnected world using IoT [4], and the authors in Ref. [4] add that such a linked system must connect smart cities, smart homes, and IoT in one cohesive paradigm. 5G technology will not only offer high-speed broadband internet connectivity, but also support e-payments, e-transactions, and other fast electronic transactions [5]. Moreover, 5G focuses on voice over internet protocol devices, and therefore, providing high level of data transmission and call volume [5].

In this chapter, we work on maximizing 5G coverage using unmanned aerial surveillance vehicles (UAVs) in urban settlements as shown in Figure 3.1. Wireless users expect to have unlimited and affordable internet access all the times. Increasing the number of base stations (BSs) in a given area is a potential way of satisfying users and to provide extended 5G coverage. However, this is not an easy task. Because a few of these BSs can have light or no load at all at a particular time, other BSs might experience very high data traffic and unnecessary overhead. The unpredictability characteristic of the user makes it hard to know exactly where and when a BS

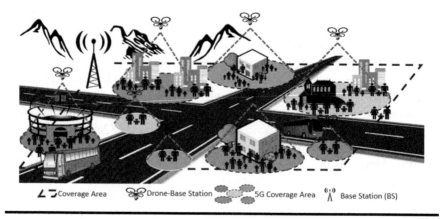

Figure 3.1 UAV-based 5G coverage in urban areas.

should be located. We use UAVs to counter this problem by designing drone-BSs as depicted in Figure 3.1. The drone-BS is flexible, able to be placed where it is needed most, and at any particular time. And hence, it efficiently provides 5G coverage for the users at all times. The author in Ref. [6] states that drone-BS can be used to provide assistance to the ground BSs with high data rates as well when additional space and time is required. There is a growing number of research work being done on drone-BS in cellular networks. However, one critical challenge that has not been given much attention is finding the lowest number of drone-BS and their respective positions in a given 3D space, required to provide maximum 5G coverage with guaranteed QoS. The main contribution of this chapter is therefore to model this challenge into liner optimized mathematical problem and use simulated annealing (SA) and genetic algorithm (GA) metaheuristic algorithms to provide quick and efficient solutions to service providers. The two algorithms are used in extensive simulations, where coverage graphs are drawn, and numerical results are compared so as to determine which algorithm can provide quick and more accurate solutions.

The rest of the chapter is organized as follows. Section 3.2 talks about some of the works that has been done relating to this study. Section 3.3 discusses some of the main challenges faced by aerial sensor networks while Section 3.4 presents the model of the system. Section 3.5 discusses the findings of this study. And finally, Section 3.6 presents our conclusions and future work. In order to further assist the reader, a list of used abbreviations in this chapter and their definitions are presented in Table 3.1.

Table 3.1 Abbreviations and Definitions

Abbreviation	Description
SA	Simulated annealing
GA	Genetic algorithm
UAV	Unmanned aerial vehicle
WSN	Wireless sensor networks
MWSN	Multimedia Wireless Sensor Network
MGSAA	Modified genetic and simulated annealing algorithm
BS	Base station
ILP	Integer linear program
MDLP	Mobile drone location problems
PM	Probability of mutation
PC	Probability of crossover
SDLP	Static drone location problems

3.2 Related Work

The efficiency of optimal drone positioning has attracted a lot of interest among researchers and academicians. The author in Ref. [7] introduces a minimum cost drone location problem. In their work, Zorbas et al. use a two-dimensional terrain to find the optimal location and number of drones to observe given targets, which could be mobile or static in a given area, and the authors develop linear and nonlinear optimization equation by considering the coverage of the drones and the energy consumed.

Moreover, authors in Ref. [8] present a study where they look into recent brainstorm optimization algorithm to find the optimal position of locating static drones in a monitored area such that the coverage is maximized. The algorithm was used to solve the placement problem for both uniformly and clustery distributed targets, according to the authors, the results obtained showed that the proposed algorithm is very efficient for solving drone placement problems. Furthermore, we can use UAVs as aerial wireless BSs when cellular networks go down. This system can be used when disasters such as flood and earthquake affect the existing communication system. The authors in Ref. [9] talk about finding an optimal position for the UAVs such that the sum of time durations of uplink transmissions is maximized. Moreover, the authors prove their hypothesis by presenting detailed simulation results for the optimization problem under different cases.

Furthermore, authors in Ref. [6] present a study on the number of 3D placements of drone BSs. In this study, the authors use a heuristic algorithm to optimally place drone BS in a region with different target densities. The goal of the study is to find the minimum number of drones and their 3D placement such that all users are served. The simulation results obtained from the study showed that the proposed system can yield QoS constraint of the network.

Numerous works have been done to compare the different results obtained by different heuristic optimization algorithms. In Ref. [10], the authors compare four studies that have been done on routing and wavelength assignment with the aim of supporting the improvement of traffic-related problems. Moreover, the authors perform various simulations using the optimizing algorithms, such as SA and GA. The results obtained revealed that the optimizing algorithm produced better results compared with the other algorithms.

The authors in Ref. [11] propose a new heuristic algorithm used to test generations of data during software testing process. Modified GA and SA (MGSAA) was used to perform different experiments. Yu et al. presents the simulation results and concludes that the proposed method generates high-quality results compared with GA. In Ref. [12], the authors used GA and SA metaheuristic algorithms to optimize a topological design network and compare the results. The authors concluded that the average GA solution cost less than the average SA solution.

However, aerial sensor network faces a lot of challenges, especially in the monitoring of outdoor critical situations where the severity of the environment, such as high temperatures, heavy rains, storms, and the likes, destroy the installed aerial sensors [13].

3.3 Challenges of Aerial Sensor Network

There are a few challenges that need to be addressed when it comes to aerial sensor deployment. First, resource allocation is one of the most important aspect. Some limited resources for sensor nodes include power, memory, and communication bandwidth. Sensor nodes consume power during operation, and while activities such as sensing, data storage, and simple data analysis are power efficient, there are other functions that consume a significant amount of power, for instance, image analysis in multimedia wireless sensor networks (MWSNs). Therefore, it is imperative that efficient power consumption systems are developed; hence, research should be focused on determining the trade-offs between locally storing, communicating and processing data, and consequently developing energy-efficient paradigms.

In aerial sensing platform, most of the power is consumed during UAV propulsion, power consumed during sensing, processing, and communication is usually relatively negligible, and hence can be ignored [14]. Therefore, for efficient power consumption, one has to plan the flight path of UAV. For instance, ascending consumes more power than flying at a constant altitude [7]. Moreover, weather conditions also have a big effect on UAV's power consumption, and the sensors on the UAV can be used to send back information about direction and speed of wind during flight.

Finally, aerial WSN communication is different compared with other communication networks. When the UAVs are flying, they need to exchange data (current position, speed, direction, etc.) with each other, as observed from Figure 3.2. Individual UAVs need to exchange their information after only a few seconds.

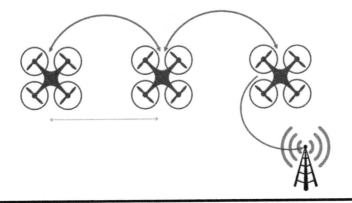

Figure 3.2   UAV communication with each other and BS.

However, multiple UAVs flying simultaneously need to know and transmit their position more accurately; hence, position data is exchanged every few milliseconds. Therefore, this system requires a communication link with low latency and a wide range of communication.

3.4 System Model

The system consists of a common sink (BS), to which information is sent and dispatched. However, since the UAVs might be out of communication range with the BS, it is efficient that the UAV sends data to the next available UAV, which in turn sends the information to the next available one, until the communication range with the BS is attained and the data can be transmitted to its final destination (i.e. BS) [13]. This system also helps to reduce the amount of energy consumed by the system since we do not require long-range transmitters that are high power consumers. Figure 3.2 demonstrates the communication paradigm of UAVs.

To obtain an optimal number of drones to be used in maximizing 5G coverage, it is imperative that drones are located in the correct position so as to obtain maximum coverage while minimizing the number of UAVs used, which in turn reduces the cost. Al-Turjman [15] says that sensor placement is a difficult but ongoing research area, considering the limited sensing and communication sensor range as well as their restricted resources such as energy and coverage (Figure 3.3).

In their study, Quaritsch et al. [14] talk about the use of UAVs in disaster management, and they discuss about the challenges facing the networked UAVs as well as focusing on their optimal placement. In doing so, Quaritsch et al. mathematically formulated the coverage problem and presented the first evaluation results. The authors took into account two optimization criteria: the first one is the quality

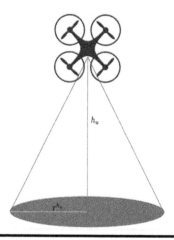

Figure 3.3 UAV cellular cover.

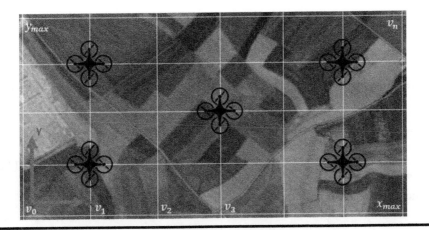

Figure 3.4   Virtual coordinates inside coverage area.

of the pictures taken, which basically refers to the coverage of UAVs, and the second is the consumption of resources, which involves the communication bandwidth and the energy used, which relates to the time of flight. The observation area and the forbidden area are drawn by the user using world coordinates, that is, longitude and latitude. However, as described by Quaritsch et al., the entire process of optimizing sensor placement is done using relative coordinates; therefore, the first step is to transform world to relative coordinates by selecting an arbitrary origin inside the observable region and formulating the x- and y-axis to go eastwards and northwards, respectively, as shown in Figure 3.4.

The authors in Ref. [7] present a study done to determine the optimal static and dynamic drone positioning in a selected area, to minimize cost and maximize coverage. The drones have a maximum and minimum observation altitude. Moreover, the height is directly proportional to the size of the observed area, as the height increases, so does the observed area as well as the energy consumed. The main focus of the study is to minimize the cost, and therefore, the number of drones or the energy consumed. The authors start with the formulation of static drone location problems (SDLP) and mobile drone location problems (MDLP).

Our drone can fly to a maximum height of h_{max} and a minimum height of h_{min}. Figure 3.4 shows a rectangle with length x_{max} and width y_{max}, which represents the area of interest. Therefore, targets could be in any arbitrary location in area $x_{max}*y_{max}$, and we assume that there is a position (x, y, h) that a drone could be located instantaneously. Let U denote a set of available drones, and T be the set of targets.

Each target $t_i \in T$ has position $\left(X_{t_i}, Y_{t_i}\right)$. Drone $u \in U$ has position (X_u, Y_u, h_u). For $h = 0$, the distance between the target and the drone is

$$D_{t_i}^{u_x, u_y} = \sqrt{\left(X_{t_i} - X_u\right)^2 + \left(Y_{t_i} - Y_u\right)^2} \tag{3.1}$$

Each drone u has a communication range θ in the form of a disc in area $x_{max}*y_{max}$, as shown by the circular area in Figure 3.3, and it has a radius of r^{h_u}, which depends on the height of the drone h_u. The bigger the value of h_u the longer the radius r^{h_u}. There are two important decisions that must be made at this point: the first one is to determine the position (X_u, Y_u, h_u) of the drone $u \in U$ (coordinates) and the second one is to find the target $t_i \in T$ in the area of interest.

For the first problem (position of drone):

$$\delta^u_{xyh} = \begin{cases} 1 & \text{if the drone } u \text{ is located at } (x, y, h) \\ 0 & \text{otherwise} \end{cases} \tag{3.2}$$

And for the second problem (target observed):

$$\gamma^u_{t_i} = \begin{cases} 1 & \text{if the target } t_i \text{ is in the vicinity of drone } u \\ 0 & \text{otherwise} \end{cases} \tag{3.3}$$

The objective is to cover all the targets using at least one drone, each drone consumes a total energy E, where

$$E = (\beta + \alpha k)t + P_{max}(k/s), \tag{3.4}$$

where β is the minimum power needed to hover at almost zero altitude, α is the motor speed multiplier, the maximum motor power is represented by P_{max}, and s and t are speed and operating time, respectively. αk represents the relation between power and height. The term $P_{max}(k/s)$ is used to show the power used to rise to height k at speed s. β and α depend on the weight of the drone and motor characteristics. Therefore,

Minimize $f(\delta)$:

Such that:

$$\sum_{(x,y,h)1} \delta^u_{xyh} \leq 1 \quad \forall u \in U \tag{3.5}$$

The drone u is located in at most one position.

$$\gamma^u_{t_i} \leq \sum_{(x,y,h)} \delta^u_{xyh} \left(\frac{r^{h_u}}{D^{u_x,u_y}_{t_i}} \right) \quad \forall u \in U, \quad t_i \in T \tag{3.6}$$

With the earlier constrain, we set the value for $\gamma^u_{t_i}$, such that, if r^{h_u} (range radius) is less than $D^{u_x,u_y}_{t_i}$ (distance), then $\gamma^u_{t_i}$ is equal to 0, in other words, if the target is outside the communication range of the 5G transmitter mounted on the drone, then the target cannot use that drone to access 5G. Therefore, the variable $\gamma^u_{t_i}$ can get either value 0 or 1.

$$\sum_{u \in U} \gamma_{t_i}^u \geq 1 \quad t_i \in \boldsymbol{T} \tag{3.7}$$

The earlier constrain ensures that there is at least one drone observing each target. The following equation shows the solution space of the decision variables.

$$\delta_{xyh}^u = \{0,1\}, \quad \forall (x, y, h), \quad 1 \leq x \leq x_{max} \tag{3.8}$$

$$1 \leq y \leq y_{max},$$

$$h_{min} \leq h \leq h_{max}, \quad u \in U \tag{3.9}$$

$$\gamma_{t_i}^u = \{0,1\}, \quad \forall t_i \in \boldsymbol{T}, \quad u \in U \tag{3.10}$$

Continuing, the function $f(\delta)$ to be minimized is of the following form:

$$f(\delta) = A - \sum_{u \in U} \delta_{xyh}^u * A_i' \tag{3.11}$$

where A is the total area to be covered and A_i' is the area covered by UAV i.
Moreover, to minimize the total energy consumed:

$$f(\delta) = \beta \sum_{(x,y,h)} \sum_{u \in U} \delta_{xyh}^u t + \alpha \sum_{(x,y,h)} \sum_{u \in U} h \delta_{xyh}^u t + \frac{p_{max}}{s} \sum_{(x,y,h)} \sum_{u \in U} h \delta_{xyh}^u \tag{3.12}$$

We propose two alternatives to solve the placement problem. GA and SA would be used to calculate the number of drones and their respective position in a given area while maintaining coverage and lifetime constrains in the 3D deployment area.

Algorithm 1: SA Pseudocode

1. Initialize: T_0, X_0, α, m, n
2. $x = X_0, x_F = X_0, T_1 = X_0$
3. For $i = 1$ to m
4. For $j = 1$ to n
5. $X_{Temp} = \sigma(x)$
6. If: $f(X_{Temp}) \leq f(x)$ then
7. $x = X_{Temp}$
8. **End If**

9. **Else if:** $U(0,1) \leq e^{-\left(\frac{f(X_{Temp}) - f(x)}{T_t}\right)}$ then

10. $x = X_{Temp}$
11. End Else if
12. If $f(x) \leq f(x_F)$ then
13. $x_F = x$
14. **End If**
15. **End For**
16. $T_{t+1} = \alpha \cdot T_t$
17. **End For**
18. **Return** x_F

Algorithm 2: GA Pseudocode

1. Initialize: PS, G_{max}, PC, PM
2. Generate initial random solutions
3. Calculate fitness for random solutions
4. Select BFS
5. **For** $g = 1$ to G_{max}
6. **For** $i = 1$ to PS/2
7. Select two parents
8. Crossover with PC
9. Mutate with PM
10. **End For**
11. Replace parents with children
12. Update BFS
13. **End For**
14. **Return** BFS

In Algorithm 1, we used SA to find the minimum number of drones such that line 1 is initializing the parameters. T_0 is the selected initial temperature in the annealing system. We use this parameter to accept or reject certain drone placement solutions. The higher the value of T_0, the higher the probability of accepting a bad solution. Hence, we start by allocating the maximum value to T_0. We gradually reduce this value using the cooling factor α, which was selected in this work as 0.95 [11]. As the temperature reduces, so does the probability of accepting bad solutions. In line 1, we also initialize the initial solution δ_0, which is heuristically selected for better results. Moreover, m is initialized for the number of stages and n for the number of movements at a given stage with a certain temperature value. The number of moves allows us to explore the neighborhood for possible efficient drone locations. Therefore, it is important that this value is carefully chosen. Line 2 assigns the initial solution to the current solution δ and to the final solution δ_F and the initial temperature to the current temperature T_1. Line 3–17 iterates over a number of stages, where we reduce the temperature of the system after every stage. While line 4–15 iterates over the number of movements at a given stage, we explore the neighboring solutions under a constant temperature. In line 5, we find a neighboring solution using the movement operator $\sigma(\delta)$, where $\sigma(\delta) = \delta + N(0,1)$. We assign this solution to a temporary solution δ_{Temp}. Line 6–8 represents an "If" statement. Line 6 checks if the temporary drone placement solution is better than the current one. To achieve this, we apply both temporary and current solutions to the fitness function shown in Eq. (3.11). Line 7 assigns the temporary solution to the current solution if the condition in line 6 is true. Line 8 ends the "If" statement. Line 9–11 covers an "Else if" statement. Line 9 uses the current temperature T_t, the temporary solution and the current solution to find an exponential value. The value is compared with a random number (between 0 and 1 exclusive) to determine whether the temporary bad solution will be accepted or not. Line 10 assigns the temporary solution to the current solution if the condition in line 9 is true. Line 11 ends the "Else if" statement. Line 12–14 represents an "If" statement. Line 12 checks if the current drone placement solution is better than the final solution using the fitness function. Line 13 assigns the current solution to the final one if the condition in line 12 is true. Line 14 ends the "If" statement while line 15 ends the second "for" loop. Line 16 computes the next stage temperature of the system T_{t+1} using the cooling factor. Line 17 ends the first "for" loop, and finally, line 18 returns the selected final solution δ_F after all iterations have been completed.

In Algorithm 2, we use GA on the same problem to find the minimum number of drones and their optimal position for the aforementioned conditions. We begin by initializing parameters in line 1. PS is the population size, representing the number of initial solutions selected. G_{max} is the maximum generation number to which an optimal solution is obtained. PC and PM are the probability of crossover and probability of mutation, respectively. These parameters are selected so as to evolve from one generation to another. In line 2, we generate initial solution in accordance with PS. This is represented by sets of 0s and 1s. In line 3, we calculate the fitness of all initial solutions, and in line 4, we select the solution with the best fitness. Line 5–10 is the iteration over generation number while line 6–13 is iteration over half the PS. We iterate over half the PS, because at every generation, we select two parents to crossover. In line 7, we select two parents, and then in line 8, we produce two children by crossover, and in line 9, we mutate the produced children using PM. In this case, we consider each element in each solution. In line 10, we end the second "for" loop, and in line 11, we replace all the parents with the newly produced children, forming the next generation of evolved drone placement solutions. In line 12, we update the best found solution (BFS) by applying the newly produced solutions to the fitness function in Eq. (3.11) and finding the best solution. This solution is compared with the previous BFS, and if it is better, we update our BFS. In line 13, we end the first "For" loop, and in line 14, we return the best solution found.

3.5 Results and Discussions

In this section, an in-depth analysis of the experimental results is presented. Java and Python were used to execute SA and GA, respectively. An area of 80 km² was selected to be observed, with each drone having a 5G transmitter with a range of at least 10 km². For SA, initialization was done as follows: an initial temperature of 300 was chosen, and an initial solution in terms 0s and 1s was chosen (1 indicating the presence of a drone in that vertex, and 0 indicating its absence).

The move operator (α) was selected as 0.95, while m and n were selected as 500 and 200, respectively. Additionally, initialization for GA was done as follows: a PS of 8 was selected, stopping criteria (i.e. G_{max}) as 50, PC of 0.5, and PM was chosen as 1.

Figure 3.5 shows the execution time for both SA and GA with a varied coverage area. In this setup, the area covered by each drone is held constant, while the total area of interest is increased from 20 to 80 km². We can observe that the execution time for both algorithms lie approximately between 0.29 and 1.2 s, with SA record-ing the fastest and GA recording the slowest time. From the graph, we also see that SA records the fastest time until the total area of interest is equal to 44 km², where both algorithms have the same execution time. However, when we increase the coverage area further, the execution time for SA slows drastically, while that of GA also slows but not as fast as that of SA. Consequently, GA realizes a faster execution time than SA for a coverage area greater than 44 km².

Therefore, we can clearly state that SA is capable of generating relevant solutions faster than GA when the coverage area is small; however, for bigger areas to be cov-ered by the UAVs, it is efficient to use GA as its time to generate optimal solutions that are much shorter than that used by SA.

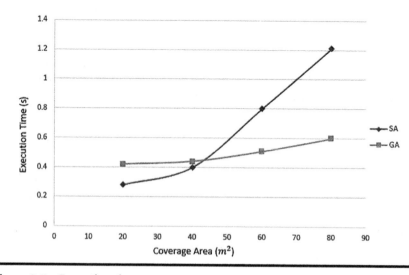

Figure 3.5 Execution time vs. coverage area.

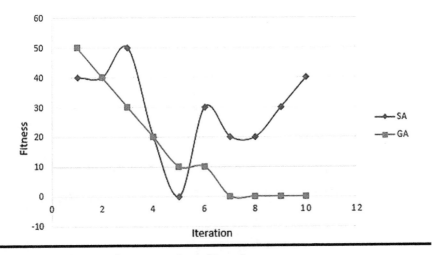

Figure 3.6   Fitness value vs. number of iterations.

In Figure 3.6, we analyze how both algorithms produce optimal solutions by tracking the fitness functions against the number of iterations. From the figure, we observe that GA consistently produces a better fitness function output than the previous one, until we get to the fifth iteration where we see a slight detonation. However, the general form of GA function depicts that the parent selection and replacement method used in our algorithm produced optimal solutions in each iteration. On the other hand, we see that SA is more unpredictable compared to GA. The algorithm has more local optimal values than GA; hence, we are more likely to get stuck on a local optimal value in SA than in GA. The figure therefore suggests that we have a high probability of getting a good optimal value when we use GA as opposed to using SA.

We generate Figure 3.7, in order to analyze the number of drones required for a given number of targets on the ground. Looking at Figure 3.7a, we observe that there are three targets, and since the targets are far from each other, we need three drones to cover all the targets. In Figure 3.7b, we increase the number of targets to ten, and here, we notice that the number of drones required increases to five. Moreover, Figure 3.7c shows that we need 11 drones when the number of targets is increased to 22. Therefore, we note that if the number of targets is x, the number of drones needed can range from one to x. Additionally, we notice that the pattern to which targets are displayed on the region of interest has a great effect to the number of drone-BS used. For instance, Figure 3.7c and d has the same number of drone-BS but has 22 and 35 targets, respectively. Moreover, Figure 3.7e and f has 81 and 111 targets respectively. But, they both have the same number of drone-BS. When targets are grouped in one area, which is within the communication range of the 5G drone-BS, we can use a single drone to cover all the targets.

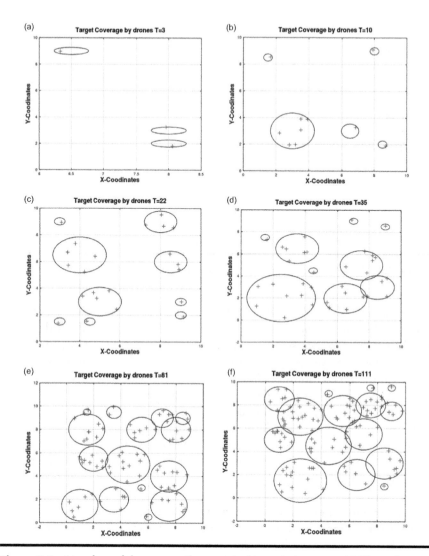

Figure 3.7   Number of drones vs. the count of targets to cover. (a) Three Drones for three targets. (b) Five Drones for ten targets. (c) Nine Drones for 22 targets. (d) Nine Drones for 35 targets. (e) Fifteen Drones for 81 targets. (f) Fifteen Drones for 111 targets.

3.6 Conclusions

In this chapter, we propose a system to determine the optimal number of drone-BS and their deployment positions needed to provide 5G cellular coverage to a given region, considering the 5G transmitter's coverage range and energy constrains of the whole system. The system is relevant when providing coverage for outdoor critical events, such as hurricane disasters, fire accidents, and densely populated areas such as urban areas and stadiums. We used two metaheuristic algorithms, SA and GA written in two different languages, to find an optimized solution. The results from both algorithms were obtained, graphed, and analyzed.

The results obtained in this study show that SA takes precedence when the coverage area is small; however, when we cover a big area, we obtain results faster using GA rather than SA. Moreover, the simulation results showed that we are more likely to settle on a local optimal value when using SA, while GA produced better results after each iteration. Generally, we conclude that using GA provides fast and better results for our outdoor UAV critical event monitoring system.

In future, we would analyze the optimal deployment problem in different environments and with dynamic UAVs. In this system, we consider making all or some of the UAVs mobile and try different patterns of UAV placement for better results.

References

1. 3GPP, Study on provision of low-cost machine-type communications (MTC) user equipments (UEs) based on LTE, June 2013.
2. M. Chen, J. Yang, Y. Hao, S. Mao, and K. Hwang, A 5G cognitive system for healthcare, *Big Data and Cognitive Computing*, vol. 1, no. 1, 2, 2017.
3. L. J. Poncha, S. Abdelhamid, S. Alturjman, E. Ever, and F. Al-Turjman, 5G in a convergent Internet of Things era: An overview, in *2018 IEEE International Conference on Communications Workshops (ICC Workshops)*, IEEE, Kansas City, MO, 2018, pp. 1–6.
4. K. E. Skouby, and P. Lynggaard, Smart home and smart city solutions enabled by 5G, IoT, AAI and CoT services, in *2014 International Conference on Contemporary Computing and Informatics (IC3I)*, IEEE, Mysore, India, 2014, pp. 874–878.
5. R. S. Sapakal, and M. S. S. Kadam, 5G mobile technology, *International Journal of Advanced Research in Computer Engineering & Technology (IJARCET)*, vol. 2, 568–571, 2013.
6. E. Kalantari, H. Yanikomeroglu, and A. Yongacoglu, On the number and 3D placement of drone base stations in wireless cellular networks, in *2016 IEEE 84th Vehicular Technology Conference (VTC-Fall)*, IEEE, Montreal, QC, 2016, pp. 1–6.
7. D. Zorbas, L. D. P. Pugliese, T. Razafindralambo, and F. Guerriero, Optimal drone placement and cost-efficient target coverage, *Journal of Network and Computer Applications*, vol. 75, 16–31, 2016.
8. E. Tuba, R. Capor-Hrosik, A. Alihodzic, and M. Tuba, Drone placement for optimal coverage by brain storm optimization algorithm, in *International Conference on Health Information Science*, Springer, Cham, 2017, pp. 167–176.
9. H. Shakhatreh, and A. Khreishah, Optimal placement of a UAV to maximize the lifetime of wireless devices. arXiv preprint arXiv:1804.02144, 2018.
10. A. Rodriguez, A. Gutierrez, L. Rivera, and L. Ramirez, RWA: Comparison of genetic algorithms and simulated annealing in dynamic traffic, in *Advanced Computer and Communication Engineering Technology*, Springer, Cham, 2015, pp. 3–14.
11. F. Al-Turjman, Mobile Couriers' Selection for the Smart-grid in Smart cities' Pervasive Sensing, *Elsevier Future Generation Computer Systems*, vol. 82, no. 1, 327–341, 2018.

12. D. R. Thompson, and G. L. Bilbro, Comparison of a genetic algorithm with a simulated annealing algorithm for the design of an ATM network, *IEEE Communications Letters*, vol. 4, no. 8, 267–269, 2000.
13. F. M. Al-Turjman, H. S. Hassanein, and M. A. Ibnkahla, Efficient deployment of wireless sensor networks targeting environment monitoring applications, *Computer Communications*, vol. 36, no. 2, 135–148, 2013.
14. M. Quaritsch, K. Kruggl, D. Wischounig-Strucl, S. Bhattacharya, M. Shah, and B. Rinner, Networked UAVs as aerial sensor network for disaster management applications, *e & i Elektrotechnik und Informationstechnik*, vol. 127, no. 3, 56–63, 2010.
15. F. Al-Turjman, Optimized hexagon-based deployment for large-scale ubiquitous sensor networks, *Springer's Journal of Network and Systems Management*, vol. 26, no. 2, 255–283, 2018.

Chapter 4

Drones Path Planning for Collaborative Data Collection in ITS

Fadi Al-Turjman and Emre Demir
Antalya Bilim University

4.1 Introduction

The revolution of cognitive computing is nowadays providing great opportunities for novel applications that promise to improve the quality of our lives. It paves the way for machines to have reasoning abilities that are analogous to human in critical scenarios. Accordingly, a new type of networks called flying ad hoc networks (FANETs) has emerged. In FANETs, the sensing and communication capabilities have been combined with unmanned aerial vehicles (UAVs) as enabling technologies. UAVs are used to physically interact with humans and urgent data collection in a shared workspace. They are flying collaborative robots that can be remotely supervised through a software-controlled flight plans in their embedded systems, working in conjunction with built-in sensors and global positioning system (GPS) modules. While regular sensor nodes passively observe their environment while attached to static objects, the ability to remotely control the UAV allocation allows us to precisely put the sensor nodes in the most suitable location. This introduces a myriad of possibilities and research challenges, especially in intelligent transportation systems (ITS). Drones (or UAVs) are widely used in safety and critical ITS applications to enhance performance and well-being. UAVs can fly over the road to monitor and observe possible traffic destructions [1–3]. For example, UAVs can fly

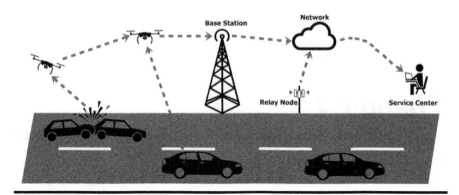

Figure 4.1   Drones can be used as flying relay node to capture video recordings of the incident scene and then relay them to a service center.

all over the place of an accident to provide urgent coverage [4]. And thus, UAVs can act like flying robots to perform several tasks as depicted in Figure 4.1.

However, the integration of existing wireless infrastructures with UAVs, especially in terms of low power with inexpensive hardware, has introduced another design perspective, which we refer to as FANET [1]. The general research challenges for designing suitable FANET to fulfill Quality of Service metrics, such as delay and data fidelity, arise primarily due to drone-limited energy budgets that must be simultaneously satisfied. Thus, we focus on optimizing the route of these UAVs in this work. Route planning options including dynamic programming should be taken into account in the application of ITS, because fast and safe paths must be assigned for data transmissions with less delay time and cost, as aforementioned. For this purpose, since dynamic programming has advantages such as calculation speed of paths, the superiority of dynamic programming in this study is implemented for finding UAVs paths and conveying data through convenient paths. Generally, roadside units (RSUs) can have easy access to battery power to work, because they are fixed objects around an infrastructure powering the network. Therefore, a kind of least cost path finder, such as the general use of simplex method and linear integer programming, can work on ITS without a doubt of energy consumption or disconnection. For instance, RSUs such as speed cameras and electronic traffic signs store and process data locally onboard due to their continuous access to power supplies. However, flying objects such as UAVs are usually powered by low-power batteries. They typically provide low data transmission and processing power, which accordingly results in a significant problem of assessing an uninterruptable ITS. This study will use drones or UAVs that are nonstatic flying objects as a contribution to the literature. Further contributions in this work can be summarized as follows. This study mainly

■ Takes into account the integration of lifetime, power, and cost of accessibility between network members employed in ITS applications.

■ Provides models that increase the speed of determining reliable transmission paths for UAVs by considering cost, communication, and power issues.

■ Optimizes the least cost path via dynamic programming to secure data routing.

■ Exhibits the superiority of dynamic programming in typical UAV network topologies for ITS, compared with other commonly used methods for the least cost path inference.

■ Demonstrates the reduction of vulnerability of data transmission by dynamic programming.

The remainder of this chapter is organized as follows. In the next section, a literature review is presented. Section 4.3 describes system models used to implement our proposed approach. In Section 4.4, we detail the proposed methodology followed by a typical use case in Section 4.5. Simulation results and discussions are outlined in Section 4.6. Finally, conclusions are provided in Section 4.7.

4.2 Literature Review

Path planning (PP) process in UAVs gained a great interest in the literature. That is due to its significant effects at the application-level performance. PP of UAVs in the internet of things (IoT) era can be achieved via several approaches. These approaches can be classified into static vs. dynamic techniques.

4.2.1 Static Approaches

As the name implies, static-based PP is dependent on a static set of data collectors (DCs) to relay and report data. For a particular engineering area, transportation engineering, the mathematical methods, and related algorithms are commonly applied to real cases [5,6]. For instance, on an UAV, transportation network of a region may need a solution to solve its travel assignment issues. At this point, network flow problems that have been used to identify the amount of flows on UAV transportation networks take place. Since the problem needs a specific mathematical structure and solution algorithms to solve, many approaches have been studied on the static approach of this transportation problem [7].

As the formulation of the transportation problem has been first discussed [8,9], the algorithm of the solution was close to the general-used simplex method [7,10,11]. The problem was applied and still being applied to many engineering optimization processes. For example, while some research proved that the transportation problem can solve transportation engineering cases of flow, such as logistics ways and times [12–14], some other research demonstrated that directing the services on a residential or commercial network can be done by the transportation problem, which were discussed using static nodes of network topologies [15–17].

A wide range of the studies of transportation problem takes place in determining the least cost path. In this case, the least cost route problems are solved in transportation-related issues by manipulating existing mathematical algorithms [7,18,19]. One of the most popular practices is to assign the least cost path by assuming all of the nodes (UAVs) in a network are at a fixed location [5,20]. Those studies used linear-integer mathematical procedures.

4.2.2 Dynamic Approaches

Approaches under this category use a dynamic set of DCs to collect and gather the information from distributed access points (APs).

To precisely gather information about a given locale, it is instructed to utilize a number of versatile DCs (e.g., Drones) and successfully use their way of intrigue. These DCs are generally navigated by low-control batteries. Subsequently, the DCs in this structure will have low information transmission control, low memory limit, and low handling force, which therefore prompts a noteworthy issue looked by any solid ITS framework. As the primary defy is interfacing with the principle base station (BS) consistently, versatile DCs were proposed in Refs. [21–23], which brought together information and choices made at the BS. Portable sensor hubs move along a predefined way in the detecting condition. There is a demonstrated advantage of utilizing DCs (versatile relays) over the traditional static sensors and/ or RSUs. A system that utilizes the previous hubs has more lifetime than a system that utilizes the last ones. Authors in Refs. [24,25] express that DCs were first used to draw out the system lifetime. The lifetime of the system is separated into equivalent amounts of time called rounds. In light of a concentrated calculation running at the BS, DCs are set toward the start of each round. The principle objective was to limit the energy of vitality utilized amid one round. It was inferred that the ideal areas, as indicated by this target work stay ideal, even after the goal is changed to limiting the greatest vitality expended per sensor. Nevertheless, utilizing this vitality metric to locate the ideal position of versatile DCs is not the best method, on the grounds that the arrangement won't be advanced regarding time. Along these lines, the DCs area found may be a long way from the ideal positions in an ITS application.

In this work, we propose an approach of dynamic programming to overcome several problems related to data gathering and conveying through a convenient path in the literature of transportation activities. For instance, in UAV transportation-related issues, many parameters including the cost (i.e. distance) between an origin and a destination location may be exposed to changes very often. In such nonstatic cases, the dynamic programming, which is one of the best way to find the least cost routes on a transportation network, can be confidently applied [7,26]. Since the dynamic programming is an approach to discover an optimal solution in large-scale algorithmic problems by dividing the problem into smaller pieces, it is an efficient way to solve relatively large-scale UAV transportation problems.

4.3 System Models

Considered FANET models are discussed in this section. A graph-based topology is assumed to model the network, and mathematically represent its connectivity. In addition, we detail the considered power and lifetime model, cost, and wireless communication assumptions. In these particular system models, the following details are accounted for this research.

4.3.1 FANET Model

In our study, the system is structured based on a communication network in a topology. We assume a network of UAVs, called FANET, connected to the cloud for remote analysis and actuation. First, the assumed FANET topology is represented by a network of G consisting of network nodes of V and the arcs of E connecting the network nodes of V. Thus, the network topology of $G(V, E)$ can denote the initial inputs of the approach. Second, the nodes V contain the information of location that is denoted spatially. Every single node has a particular latitude and longitude information. For $v(x, y)$, where x is for the latitude data, y is the longitude data. And V consists of all v's, which are all included in the network topology G. The arcs E represent the cost of connecting the nodes V. Here the cost means not only the monetary cost of connectivity but also the distance or time. In typical networks, there are generally one origin node and one destination node. However, in our study, all of the network nodes of V, except one destination node v_d, are the starting nodes.

4.3.2 Cost and Communication Models

Since UAVs are the most expensive component in the assumed FANET, cost is modeled in this chapter by the count of UAVs hovering in the targeted site. For simplicity, we consider similar cost for utilized UAVs. As for the wireless communication, we assume a probabilistic model as described in Eq. (4.1), where remote signals are decayed by separation distances, and influenced by surrounding obstacles [27].

$$P_r = K_0 - 10\gamma \log(d) - \mu d \tag{4.1}$$

where P_r is the gotten signal strength, d is the Euclidian separation between sender and recipient, γ is the path loss, μ is an irregular variable that pursues a log-normal distribution with zero mean and difference δ^2 to depict signal lessening impacts in the observed segment, and K_0 is a consistent dependent on the sender, recipient, and observed region heights. This wireless communication model applies for inter- and intra-FANET connections, where a cellular network is assumed to connect a sink UAV to the cloud server and a ZigBee-based connection for the intercommunication between UAVs (or the FANET nodes) as depicted in Figure 4.2.

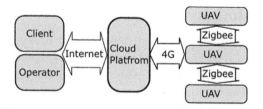

Figure 4.2 Connectivity diagram.

4.3.3 Power and Lifetime Model

In addition to energy consumed by communication, energy (power) can be consumed in an UAV in hovering and hardware components. We assume that all utilized UAVs have similar characteristics. These UAVs move from the starting point to a given destination at a speed represented by v_u. Hovering and hardware consumed power, indicated by P_{hov} and P_{har}, are represented by [28]:

$$P_{hov} = \sqrt{\frac{(m_{tot} g)^3}{2\pi r_p^2 n_p \rho}}, \text{and } P_{har} = \frac{P_{full} - P_s}{v_{max}} v_u + P_s \qquad (4.2)$$

where m_{tot}, g, and ρ are the UAV mass (in kg), earth gravity (in m/s²), and air density (in kg/m³), respectively. r_p and n_p are the radius and count of UAV fans. v_{max} is the highest achievable speed by the UAV. P_{full} and P_s are the hardware power consumptions when the UAV is moving at the highest speed and when the UAV is in a steady state. Given the aforementioned power consumption models and the initial energy budget per UAV, we can estimate the amount of time each UAV can stay alive (functional). Hence, we assume the following lifetime definition:

Definition 1

FANET lifetime is defined by the amount of time period since the deployment of UAVs till the time a percentage of UAVs that are connected reach a minimum predefined threshold τ.

4.4 Least Cost Path Finder (LCPF) Approach

In this section, we explain the details of our proposed approach in finding the most appropriate UAV path in terms of cost. Dynamic programming is applied to overcome the aforementioned problems related to data gathering by finding the most convenient UAV path. According to Refs. [27–29], the formulation of dynamic programming can be achieved as follows. The decision variables x_n are the immediate destination, where n represents the particular stage (e.g. $n = 1, 2, 3, \ldots, N$). Therefore, $A \rightarrow x_1 \rightarrow x_2 \rightarrow x_3 \rightarrow \ldots \rightarrow x_n$ becomes the route chosen, where A is the main origin and $x_n = B$ is the final destination.

For the best overall selection for the rest of the stages, $f_n(s, x_n)$ can depict that the seeker at node s, just about to start stage n, can choose x_n as an immediate destination. By taking into account s and n, x_n^* denotes a value of x_n that minimizes $f_n(s, x_n)$. In addition, the minimum value of $f_n(s, x_n)$ is shown by $f_n^*(s)$. Therefore, the following equations can be sorted out.

$$f_n^*(s) = \min_{x_n} f_n(s, x_n) = f_n\left(s, x_n^*\right) \tag{4.3}$$

where

$$f_n(s, x_n) = (\text{immediate cost stage } n) + \text{minimum future cost (stages } n + 1 \text{ onward})$$

$$= c_{sx_n} + f_{n+1}^*(x_n) \tag{4.4}$$

$$\sum P_{hov} + \sum P_{har} \le E_t \tag{4.5}$$

$$\sum UAV_{rounds} \le \tau. \tag{4.6}$$

where c_{sx_n} represents the cost between the current node and the immediate destination. This cost is calculated based on total traveling distance and overall collected data. In dynamic programming, besides, the ultimate goal is to find $f_1^*(s)$ and the route corresponds to it, and the other targets are to determine the routes of $f_2^*(s)$, $f_3^*(s), \ldots, f_n^*(s)$ sequentially.

In this approach, the crucial interest is to identify the least cost path from the first origin node (i.e. the node-A) to the final destination (i.e. node-B). However, since there might be introductions for data joining the network from the internal stages (e.g. $n = 2, 3, \ldots, n-1$), the least cost paths heading to the final destination from those nodes are also in concern. Therefore, in summary, our algorithmic steps can be described as follows.

Algorithm 1: Least Cost Path Finder (LCPF)

Begin:
1. *Stage-1:* Looking for an answer of which node should be used to arrive at the final destination as an immediate destination along the least cost path.
2. *Stage-2:* Using the results of *Stage-1*, *Stage-2* looks for an answer of which node should be used to arrive at the final destination node along the least cost path.

3. **Stage-3:** Using the results of *Stage-2*, *Stage-3* looks for an answer of which node should be used to arrive at the final destination node along the least cost path.

4. ... (The search goes on until the last stage, which is the determination of the way just after the main origin node or the immediate destination after the main origin).

5. **Stage-n:** Using the results of *Stage-(n−1)*, *Stage-n* looks for an answer of which node should be used to arrive at the final destination node along the least cost path.

End.

Algorithm 1 is the least cost path finder (LCPF) where *Stage-1* represents the initial path discovery for arriving the final destination node, which is BS1 as an immediate destination (i.e. the next adjacent destination) through the least cost path. Next, *Stage-2* denotes the path discovery using the results of *Stage-1*. It seeks an answer of which immediate network node (i.e. the next adjacent node) should be visited to arrive at the final destination node along the least cost route. Likewise, using the results of *Stage-2*, *Stage-3* seeks a network node to select as an immediate destination to be visited, for heading to the final destination node along the least cost route. The search for the nodes (i.e. the immediate nodes that are on the way of the least cost route) continues until the last stage (i.e. *Stage-n*), which is the determination of the direction just after the main origin node or the immediate destination after the main origin. Assuming there are totally n stages for determining the immediate nodes to be visited along the least cost path, using the results of *Stage-(n−1)*, *Stage-n* looks for an answer of which network node should be visited to arrive at the final destination node along the least cost path. As in Algorithm 1, which defines the LCPF, the problem to find the least cost route is divided into smaller ones or questions to solve a large size problem in a network topology.

4.5 Use Case

In this section, to demonstrate the methodology and ability of the dynamic programming with its application steps to find the optimal solution, a case study is introduced. We expound more on our proposed methodology through a commonplace deferral-tolerant correspondence situation in ITS. We delineate a precedent where three locales in a city have the APs: A, B, and C, which have set up at least one interfacing ways, through a few DCs, to the BS1 appeared in Figure 4.3. Note that every DC way has its start-to-destination stockpiling limit and voyaged separation attributes, as conjectured by the proposed methodology given in the form (limit/distance) in Figure 4.3. These qualities depend on directing table trades among DCs and APs, which is accomplished by means of Eqs. (4.3) and (4.4) that adopt these characteristics. The proposed method goes for choosing the ways

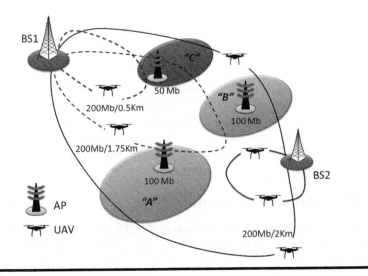

Figure 4.3   A use case for APs of a group of regions serving multiple BSs.

(appeared dashed lines) that ensures the least all-out voyaged distance without disregarding the DC-limited resources. The choice of using these DCs relies upon its assets which are put away, too, at the AP's directing table. For instance, accepting that AP in area A in Figure 4.3 needs to transmit data packets with an information stack that requires a base DC limit of 100 MB, the proposed dynamic programming approach detects a couple of ways interfacing district A to the BS. Every one of them declares its limitations, and the one among these two, which gives the base all-out voyaged distance, is eventually picked. It will be additionally ready to serve the AP at locale B as the rest of the capacity of that DC (or UAV) is as yet fulfilling its limitation. And after that, the second course (dashed line) going by area C will be used to convey its information. Furthermore, subsequently, the chose ways will be

Path 1: visiting the nodes in the sequence of *BS1*, region-*A*, and region-*B*.
Path 2: visiting the nodes in the sequence of *BS1*, region-*C*.

Note that the nondashed way fulfills the DC requirement, notwithstanding, it doesn't give the base-voyaged separation altogether on the off chance that it has been decided for locale A and B. The choice of the base DC check in the nearness of each AP in the system is accomplished by occasionally trading steering tables or potential enlistment records with BSs to convey delay-tolerant information data packets. This choice procedure is rehashed toward the start of each activated round as we previously mentioned. The versatility history of DCs is inspected against the correspondence scope of the comparing goal (AP). In light of the outcomes, DCs are characterized by best, for example, least voyaged distance altogether per round.

Among the most critical preparing suggestions is the portability factor. Since the normal experienced speed per DC in the expected Drone-based setup is between 15 and 30 km/h, there will be sufficient time to trade messages utilizing the IEEE 802.15 standard. In any case, the collected data traffic can be either real-time or nonreal-time for more client fulfillment. In the meantime, traffic amount, speed of portable DCs, and the multiinterface DC are for the most part essential parameters to accomplish the best vitality utilization by the ITS framework. For example, expanding the DCs' request for rate in the shrewd city can prompt an augmentation in the holding up time and vitality utilization per information unit.

4.6 Performance Evaluation

To further investigate and assess the proposed LCPF approach using dynamic programming, the data in the case study were processed in LINGO 17.0 x64. The experiment setup in the case totally contains 14 DCs as the nodes of the network, which is created by a random topology for these DCs (Figure 4.4). The cost between the DCs is valued by distance in kilometers. The script as well as the data were entered into the software [30]. The solver was run by a processor with the properties of Intel® Core™ i7–2640M CPU at 2.80 GHz. As the program was run by evaluating the inputs and completing the optimization, the least cost path was determined as follows.

DC-1 → DC-2 → DC-4 → DC-9 → DC-14 → BS1

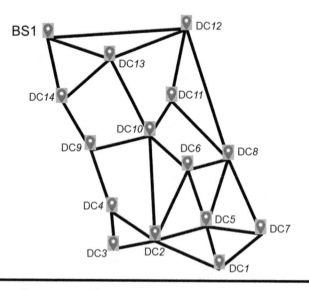

Figure 4.4 **Totally, 14 DC nodes and BS1 at the network are created by a random topology.**

The path starts from DC-1 as the origin node and terminates at the destination node of 15 at BS1. In other words, one should be visiting the DCs of 1, 2, 4, 9, 14, and node-15 (i.e. BS1), respectively, to access node-15 (i.e. BS1) from DC-1 with the help of the least cost path.

Assuming all paths are feasible by the solver, the least cost routes with their costs (i.e. distances) from each DC node to the final destination node BS1 were also successfully provided, as shown in Table 4.1. For instance, the least cost route distance in the network from an origin of DC-4 to the destination node-15 (i.e. BS1) has 109 units of cost totally. Likewise, the least cost path from DC-9 as a starting point to the destination of BS1 requires 71 units of cost. Clearly indicated in Table 4.1 that the least cost path distance from node-15 (i.e. BS1) to node-15 (i.e. BS1) should be zero, as computed well by the solver.

The following figures (i.e. Figures 4.5–4.7) are constructed to compare the dynamic programming method with a couple of other commonly used methods, which are the least cost path algorithm by Dijkstra [18] and minimal spanning tree (MST) methods [31,32]. In this manner, several networks are used as inputs to find the results. Since the typical number of drones or UAVs on a city is approximately 35, the number of drones, UAVs, or nodes in the network topology has been chosen up to 35 nodes. Therefore, data of a couple of networks including 10, 15, 20, 30, 33, and 35 nodes are processed to reveal the performance of dynamic programming against the other generally used methodologies.

In Figure 4.5, the average cost between the nodes, which is the distance in kilometers in this study vs. the number of nodes in the network topology, is depicted. The paths inferred by the dynamic programming methods and the Dijkstra's algorithm are exactly the same in those cases. In other words, the paths found by two methods are matching in terms of the nodes visited. Not only they are similar but also they are the least cost ways from the origin and the destination at the particular network topology. However, the distances of paths created by the MST method are always more than the other two. For example, the path created by dynamic programming and Dijkstra's methods has a length of almost 400 km in

Table 4.1 The Costs of the Least Cost Paths from the Origins to the Main Destination Node BS1

From	Distance	From	Distance	From	Distance
DC-1	185	DC-6	124	DC-11	127
DC-2	141	DC-7	195	DC-12	85
DC-3	131	DC-8	166	DC-13	41
DC-4	109	DC-9	71	DC-14	39
DC-5	160	DC-10	93	BS1	0

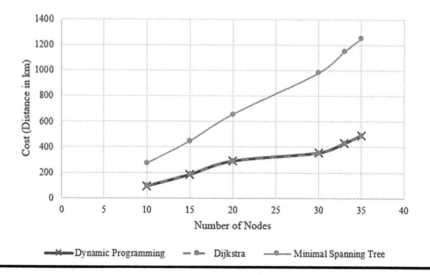

Figure 4.5   Average cost (distance) vs. number of nodes in the network topology.

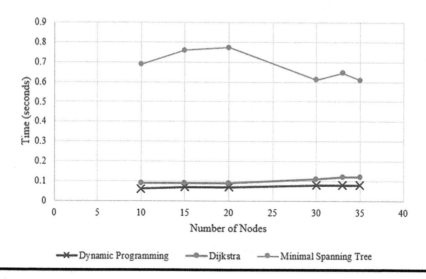

Figure 4.6   Time to find the result in seconds vs. number of nodes in the network topology.

the network of 30-node, while the path found by MST has a distance of almost 1,000 km (Figure 4.5). Moreover, as can be seen in Figure 4.5, the gap on the graph between the dynamic programming and Dijkstra's method gets larger as the number of nodes increases.

Figure 4.6 demonstrates time in seconds to find the results; in other words, the computer processing time in seconds to set the results is very less for the

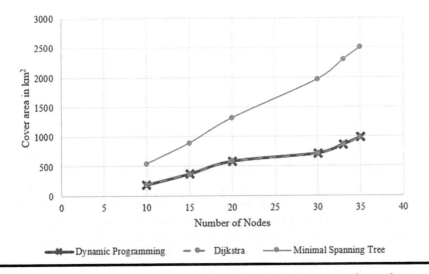

Figure 4.7   Average cover area vs. number of nodes in the network topology.

methodologies of dynamic programming and Dijkstra. Although the time for both is less, dynamic programming is faster to capture the results and provide the solutions than the processing time of Dijkstra's algorithm (Figure 4.6). The processing data using the MST method is unfortunately time consuming compared to the methods of dynamic programming and Dijkstra. Even though the time to find the result slightly decreases as the number of nodes increases for the method of MST, the difference of processing time is still really much between the methods of dynamic programming and Dijkstra and the method of MST. Furthermore, as the number of nodes increases, while the processing time of Dijkstra algorithm slightly increases, the time to find results using dynamic programming is almost steady.

In addition, Figure 4.7 shows the cover areas by the paths delivered by three methodologies. Since drones or UAVs fly over the city and transmit the data between each other, the transmit paths surely have a typical coverage area that is roughly 2 km on the land to be detected. Detection may cause negative effects such as disturbance of the data transmission; therefore, the shorter path is better to avoid unexpected trouble on transfer data. For instance, if the length of the least cost path is almost 100 km for the 10-node network, the cover area of that path can be about 200 km^2 (Figure 4.7). Therefore, this parameter depends on the length of the path inferred, and there is a direct relationship between the length of path distance and the cover area. Paths created by dynamic programming and Dijkstra's methods are similar to one another, but the areas of paths offered by MST are more than the other two methods. Although the cover area by the MST method is larger than the paths of the other two methods, this is because the length of the paths by MST method are longer, which makes the travel of the signals slower and more probable to be exposed to disturbance in comparison. According to the results obtained, overall, the performance of dynamic programming is better than the other commonly used methodologies in terms of finding time of the results and also inferring the least cost paths for the network topologies to transmit fast and safe data.

4.7 Conclusions

In this chapter, by applying mathematical techniques and models, several problems related to ITS and its conveying data techniques come up with solutions including the assignment of least cost paths according to the structure of a network topology. The least cost transportation routes were determined by a commonly used methodology of dynamic programming, considering the parameters of a network in cases given. The contribution of this chapter is not only in making smart decisions on selecting the UAV transportation ways in nonstatic phases of a massive regional data traffic, but also in reducing the total needed UAV count by utilizing the most overlapping routes.

A future research could focus on not only controlling the determination of paths but also affording an integration of data transferring or conveying in a combination of a couple of city networks. Relatively large datasets could be required and processed for facing challenging effects of complicated network topologies of UAV use.

References

1. F. Al-Turjman, E. Ever, and H. Zahmatkesh, Small cells in the forthcoming 5G/IoT: Traffic modelling and deployment overview, *IEEE Communications Surveys and Tutorials*, 2018. doi:10.1109/COMST.2018.2864779.
2. L. Selçuk, An avalanche hazard model for Bitlis Province, Turkey, using GIS based multicriteria decision analysis, *Turkish Journal of Earth Sciences*, vol. 22, 523–535, 2013.
3. F. Al-Turjman, H. Hassanein, and M. Ibnkahla, Towards prolonged lifetime for deployed WSNs in outdoor environment monitoring, Elsevier Ad Hoc Networks Journal, vol. 24, no. A, 172–185, January 2015.
4. F. Al-Turjman, Energy–aware data delivery framework for safety-oriented mobile IoT, IEEE Sensors Journal, vol. 18, no. 1, 470–478, 2017.
5. T. C. Matisziw, and E. Demir, Inferring network paths from point observations, *International Journal of Geographical Information Science*, vol. 26, no. 10, 1979–1996, 2012.
6. E. Demir, Assigning convenient paths by an approach of dynamic programming, Dedicated to Professor Gradimir V. Milovanović on the Occasion of his 70th Anniversary, MICOPAM 2018, pp. 196–199, 2018.
7. F. Al-Turjman, Cognitive routing protocol for disaster-inspired internet of things, Elsevier Future Generation Computer Systems, vol. 92, 1103–1115, 2019.

8. F. L. Hitchcock, The distribution of a product from several sources to numerous localities, *Journal of Mathematical Physics*, vol. 20, 224–230, 1941.
9. D. R. Fulkerson, Hitchcock transportation problem. No. P-890. Rand Corp, Santa Monica, CA, 1956.
10. R. K. Ahuja, T. L. Magnanti, and J. B. Orlin, *Network Flows*, Prentice-Hall, Englewood Cliffs, 1993.
11. T. Tokuyama, and J. Nakano, Efficient algorithms for the Hitchcock transportation problem, *SIAM Journal on Computing*, vol. 24, 563–578, 1995.
12. U. Brenner, A faster polynomial algorithm for the unbalanced Hitchcock transportation problem, *Operations Research Letters*, vol. 36, 4, 408–413, 2008.
13. A. Sharma, V. Verma, P. Kaur, and K. Dahiya, An iterative algorithm for two level hierarchical time minimization transportation problem, *European Journal of Operational Research*, vol. 246, 3, 700–707, 2015.
14. F. Al-Turjman, and S. Alturjman, 5G/IoT-Enabled UAVs for multimedia delivery in industry-oriented applications, Springer's Multimedia Tools and Applications Journal, 2018. doi:10.1007/s11042-018-6288-7.
15. F. Al-Turjman, M. Z. Hasan, and H. Al-Rizzo, Task scheduling in cloud-based survivability applications using swarm optimization in IoT, Transactions on Emerging Telecommunications, 2018. doi:10.1002/ett.3539.
16. R. A. Maher, and A. F. Alrouby, An algorithm for cost-minimizing in transportation via road networks problem, *International Journal of Mathematical and Computational Methods*, vol. 2, 292–299, 2017.
17. E. Demir, Havalimanlarında kalkış öncesi, acil durumlarda, yardım alınabilecek en uygun lokasyonun Weber problemine uyarlanarak belirlenmesi, *Türk Coğrafya Dergisi*, vol. 70, 81–85, 2018.
18. E. W. Dijkstra, A note on two problems in connection with graphs, *Numerische Mathematik*, vol. 1, 269–271, 1959.
19. L. R. Ford Jr., *Network Flow Theory*, Rand Corp, Santa Monica, CA, 1956.
20. J. Current, C. ReVelle, and J. Cohon, Symposium on location problems: In memory of Leon Cooper: The shortest covering path problem: An application of locational constraints to network design, *Journal of Regional Science*, vol. 24, 161–183, 1984.
21. F. Al-Turjman, A. Alfagih, H. Hassanein, and M. Ibnkahla, Deploying fault-tolerant grid-based wireless sensor networks for environmental applications, in *Proceedings of the IEEE International Workshop on Wireless Local Networks (WLN)*, Denver, CO, 2010, pp. 731–738.
22. F. Al-Turjman, H. Hassanein, S. Oteafy, and W. Alsalih, Towards augmenting federated wireless sensor networks in forestry applications, *Personal and Ubiquitous Computing*, vol. 17, no. 5, 1025–1034, June 2013.
23. G. Solmaz, M. I. Akbas, and D. Turgut, A mobility model of theme park visitors, *IEEE Transactions on Mobile Computing (TMC)*, vol, 14, no. 12, 2406–2418, 2015.
24. F. Al-Turjman, H. Hassanein, and S. Oteafy, Towards augmenting federated wireless sensor networks, in *Proceedings of the IEEE International Conference on Ambient Systems, Networks and Technologies (ANT)*, Niagara Falls, ON, Canada, 2011, pp. 224–231.
25. L. Bloni, D. Turgut, S. Basagni, and C. Petrioli, Scheduling data transmissions of underwater sensor nodes for maximizing value of information, in *Proceedings of IEEE GLOBECOM'13*, December 2013, pp. 460–465.

26. R. Bellman, On the theory of dynamic programming, *Proceedings of the National Academy of Sciences*, vol. 38, 8, 716–719, 1952.
27. F. Al-Turjman, Price-based data delivery framework for dynamic and pervasive IoT, Elsevier Pervasive and Mobile Computing Journal, vol. 42, 299–316, 2017.
28. F. Al-Turjman, A novel approach for drones positioning in mission critical applications, Wiley Transactions on Emerging Telecommunications Technologies, 2019. doi:10.1002/ett.3603.
29. A. Alchihabi, A. Dervis, E. Ever, and F. Al-Turjman, A generic framework for optimizing performance metrics by tuning parameters of clustering protocols in WSNs, Springer Wireless Networks, vol. 25, no. 3, 1031–1046, 2019.
30. LINDO Systems Inc. (2018). www.lindo.com/index.php/products/lingo-and-optimization-modeling. Last access September 1, 2018.
31. J. B. Kruskal, On the shortest spanning subtree of a graph and the traveling salesman problem, *Proceedings of the American Mathematical Society*, vol. 7, 1, 48–50, 1956.
32. H. Loberman, and A. Weinberger, Formal procedures for connecting terminals with a minimum total wire length, *Journal of the ACM (JACM)*, vol. 4, 4, 428–437, 1957.

Chapter 5

5G/IoT-enabled UAVs for Multimedia Delivery

Fadi Al-Turjman and Sinem Alturjman

Antalya Bilim University

5.1 Introduction

Advances in the internet of things (IoT) technologies and some new emerging information and communication technologies (ICT), such as the 5G devices/sensors, are converging with a variety of application fields [1,2]. Its integration with the industry is envisaged to revolutionize the current industry by applying smarter machines, building connectivity between them, allowing them to communicate with and control one another for collaborative automation and intelligent optimization. 5G is expected to be more than a new generation of mobile communications [3]. Instead, it is already considered as the unifying fabric that will connect billions of devices in some of the fastest, most reliable, and most efficient ways possible. Of course, the impact of such an enabling technology is expected to be revolutionary. The new infrastructure for communication is expected to transform the world of connected sensors and reshape industries. Such a revolution would of course require research and development for the coexistence and device interoperability for sensors with 5G networks.

Drones, also known as unmanned aerial vehicles (UAVs), have been used mainly in military applications for many years. However, there has been a recent increase in the use of UAVs in nonmilitary fields, which is inspired by the 5G revolution. Such fields include precision agriculture, security and surveillance, delivery of goods, and provisioned services [1]. For example, Amazon and Walmart have been working on a new system to deliver goods to customers over the air. Additionally,

China's largest mailing company, DHL, has started delivering around 500 parcels daily using UAVs. Moreover, we can use some 5G-supported UAVs to monitor and send feedback from incidents that happen along the road, hence, eliminating road support teams. Moreover, a traffic policeman can be replaced or assisted by a UAV, by hovering over fast-moving vehicles and report back traffic violations. Consequently, the use of UAVs for industry-oriented services can become a reality very soon, especially after the revolution of communication systems towards realizing the 5G-inspired IoT (5G/IoT) paradigm, where a key field of interest for IoT and sensor networks is the development of wearables that can connect to these UAVs for various application areas. Having an infrastructure such as 5G, which is developed considering IoT applications in detail, causes significant needs for further contributions in terms of data, especially multimedia delivery. 5G is targeting 10 Gbps data rates in real-time networks [4]. The recent tests in 5G Innovation Centre [5] have shown that it is even possible to exceed 1 Tbps in laboratory environments. This would mean being able to transmit 33 high definition (HD) films each up to 2.5–3 h in 1 s. It is typically desirable to have these infrastructures seamlessly integrated with IoT industrial solutions, and there are recommended prototype sensors for similar applications with the energy and marginal cost for each added sensor [6].

Wireless sensor networks (WSNs) are very critical in the aforementioned archetype. The integrated 5G and IoT are termed as an extraordinary complex model, where devices are deployed as consumer elements forming a complex interconnected system. Conversely, these elements operate with very strict energy constraints, and hence making the energy left over for fault-tolerance procedure limited. Moreover, the emergence of the variety of multimedia IoT applications, such as video streaming from smart homes in smart cities, will certainly increase the need for fault-tolerant data routing [7].

Nowadays, WSNs function in an autonomous manner with very limited human control in a UAV-enabled system, where sensors and cameras are attached/distributed not only in smart environments but also to flying UAVs in the industry. Moreover, most of these sensors are positioned in wild outdoor environment and sometimes even harsh environment. Hence, it is quite difficult to determine and design a fault-tolerant routing protocol. Because the communication energy is considerably lower than that used in computations, it is very important to come up with a fault-tolerant algorithm that is able to recover from path failures no matter the added computational energy is. Or else, any random event may cause UAV failure in delivering their exchanged information and interrupt the network functionality.

Accordingly, this necessitates a multipath routing approach that can recover the failed path. Multipath routing protocols form a good candidate for more reliable 5G/IoT paradigm, in which fault-tolerance routing problems are considered as optimization problems. These optimization problems formulate k disjoined paths to encounter up to $k - 1$ path failure. Exceptional fault-tolerance routing in UAV-enabled networks needs a huge computational power, which as the problem increases, brings about large control message overhead without scalability [8]. Coming up with a solution to these problems on each sensor may require significant capacities in terms of memory and computational resources, and still produce ordinary results.

To offer quick recovery from failures, we design a bioinspired routing algorithm called particle swarm optimization (PSO). Authors in Ref. [9] note that the use of PSO has produced positive results, due to its simple concept and high efficiency. Nonetheless, despite its competitive performance, there still remains a huge challenge of solving fault-routing problem because of the convergence issue. However, many of the impulsive convergence traps occur due to fast convergence features and a diverse loss of particle swarm, and hence, result into different solutions. In addition, the ability to differentiate between exploration and exploitation search is another significant challenge that we face today. Exploration contains the swarm convergence, while exploitation usually tend to make the swarm particle convergence without leaving the viable area that eventually leads to premature convergence, hence it is never proper to overemphasize exploitation or exploration [10]. Due to these faced challenges, and especially connectivity issues, we propose a new approach that is more efficient in recovering failures via multipath routing capable of attaining quality of service (QoS) in terms of energy consumption, lifetime, delay, and throughput. The proposed multipath routing algorithm is compared against existing optimization algorithm, namely canonical particle swarm (CPS) [11], fully particle multipath swarm (FMPS) [12], and multiswarm PSO (CPMS) algorithm [10] to offer a different learning technique for swarm particles. The aforementioned algorithms are different from each other in that they have different learning contrivances and the likes; otherwise, they are similar to each other. Additionally, increasing the number of paths requires more messages exchange and communication overhead [11]. Therefore, we adopt the use of intricate network connection so as to denote the layout of the swarm and use the multipath routing algorithm to stabilize the trade-off between fault tolerance and communication overheard by taking advantage of a mixed proactive and reactive routing mechanism that maintains the best objective function value for the designated path per particle. After which, the particles are increased or decreased then, given a velocity that suits them, the augmented objective function must be used to make a fitting assortment.

In view of industrial IoT (IIoT) solutions for manufacturing industry from system and network perspective, this study endeavors to provide novel data delivery solutions to gain machine/sensors interoperability and manufacturing flexibility through production line level machine collaborations, focusing on (1) sensor/machine functionality and decentralized structure for communication-intensive applications; (2) ubiquitous message trading and learning techniques for collaborative automations; and (3) swarm-based management for application-level flexibility and adaptation. Due to the aforementioned issues in WSN technology, especially the connectivity ones, we propose a new routing algorithm that is more efficient in considering multipath failures that contain reconstructive procedures capable of attaining QoS in terms of network lifetime, energy consumption, delay, and throughput.

5.2 Related Work

IIoT, also known as industrial internet, brings together smart machines, innovative analytics, and people at work. It is an interconnection of many devices through a diverse communication system to bring forth a top-notch system capable of monitoring, collecting, exchanging, analyzing, and delivering valuable information. These systems can then help manage smarter and faster business resolutions for industrial companies. This futuristic concept is marked with some new coined terms by industrial professionals and communities, such as Industrial 4.0, IIoT, Smart Manufacturing, Digital Manufacturing, Manufacturing 2.0, and Industrial Internet.

IIoT is more advanced than commercial IoT, simply due to the dominance of the connected sensors in an industrial platform. Sensor interface is a key factor in industrial data collection, and the present connected number, the rate of sampling, and the type of signals emitted by sensors are determined by the sensing device [13]. Additionally, every sensor connected to a device needs to write long and complex codes. Hence, the author in Ref. [13] proposes a different system to design a programmable sensor interface for IIoT WSN, which is able to collect data and at the same time read it in real time from multiple sensors in high speed. There are many motivations associated with IIoT, such as connecting sensors to analytics and other data processing systems to automatically improve the industrial system performance, safety, reliability, and energy efficiency while collecting data using sensors, which have proved to be effective in terms of cost [14,15]. The performance of machines and their connectivity, communication, and data throughput are all expected to be more flexible or powerful in manufacturing industry compared with traditional scenarios, such as smart home, health monitoring, and elderly care. The specificities of IIoT manufacturing industrial networks can be briefed as follows: (1) powerful machines connected to sensors with reliable connectivity, (2) real-time communication for collaborative automation, (3) heterogeneous networked machines topology, (4) arbitrary peer to peer (P2P) or broadcast for ubiquitous messaging, (5) high data throughput with varying sized messages using next-generation paradigms (e.g., 5G), and (6) machine collaboration-based intelligent automation. These features have raised challenges and become the major concerns in the development of data routing in IIoT industrial systems.

In general, providing reliable services for applications that demand low latency within the 5G and IoT context is a challenging issue. It is well known that some WSN industrial applications require deterministic systems with a reliable and low latency aggregation service guarantees. Since the IEEE 802.15.4e standard is considered as the backbone of WSN-enabled IoT, further contributions/attempts are required to overcome its latency issues and fulfill the major requirements in multimedia applications. In turn, research groups for this standard have studied on improvement of QoS-related concerns including energy efficiency [16]. Apart from the fact that fault-tolerant routing makes the network system more reliable, it is also very important when it comes to 5G or IoT, since WSNs heavily depend on surrounding environments and interact with, and hence, the need to provide QoS is a must. There are different ways to determine a fault-tolerant route in WSNs, and one of the leading ways is multipath routing [17]. Authors in Ref. [18] claim that recent optimization methods, including metaheuristic ones, are more effective in multipath routing. Moreover, the author in Ref. [19] comes up with a solution to the disjoined multipath problem, by proposing a new energy-efficient multipath routing system based on using PSO. The author in Ref. [20] introduces PSO, which is used to select routes for load delivery. On the other hand, the author in Ref. [21] recommends an enhanced PSO-based clustering energy optimization system that minimizes the use of power in each node by centralized clusters and optimized cluster heads. Nevertheless, none of these studies address jointly the lifetime and fault-tolerant aspects in their routing algorithms with a convergent model.

Authors in both Refs. [21,22] present a model to prevent the unnecessary convergence of crowd by setting upper and lower search space bounds so as to enable the crowd to find solutions for diverse applications. Furthermore, the author in Ref. [23] demonstrates the performance of a load distribution system so as to address the optimization problem and facilitate prime network selection. Moreover, the authors account for the network bandwidth and the errors in the ideal data delivery system in different networks with reduced cost. To improve the lifetime and increase the bandwidth of energy-proficient distributed clusters of sensors, authors in Ref. [24] employ the use of energy-aware, delay-tolerant, and centralized approach. However, traditional sensor networks spend energy in almost all processes. They spend energy while making data transmission and data sensing as well as data processing. There have been a few attempts towards achieving more energy efficiency in such networks via wireless multihop networking, such as Refs. [25,26]. However, such schemes are mostly applicable in static environments and can struggle with random topologies. For example, a routing scheme for energy harvesting was proposed in Ref. [27]. The routing scheme assumes a hierarchical cluster-based architecture. Packet transmission from the source to the cluster head can be direct or multihop based on the probability of saving energy through careful transmissions, optimized throughput, and minimized workload.

In this work, we propose a new routing algorithm that is more efficient in considering multipath failures that contain reconstructive procedures capable of attaining QoS in terms of network lifetime, energy consumption, delay, and throughput, while taking into consideration the advantageous powerful machines in IIoT industrial systems.

5.3 System Model

The routing method that has been projected in this research uses fault-tolerant system in two-tiered heterogeneous WSNs that comprise super/smart nodes that has plenty of resources and simple/light sensor nodes with limited battery capacity and absolute QoS limitations. Nevertheless, to get a more resilient fault-tolerant network model, we look for a K-disjoined multipath routing approach. Furthermore, we look at a many-to-one traffic system, where super nodes and common nodes connect with a proper degree. Below, we list some important explanations of some terms before introducing the system model.

For every disjoined/isolated node, we have to use it to build a k-disjoint multipath route and increase the number of marginal paths, rendering a fault-tolerant network. Our model is based on the assumption that a given node can connect or disconnect with some nodes that are not among those that are on the k-disjoint multipath between the node and a super node. In this study, node-disjointness relations are modeled as a directed graph $G(V, E)$, where $|V| = \{v_1, v_2, \dots, v_N, v_{N+1}, \dots, v_{N+M}\}$ is the fixed number of nodes or particles, N indicates sensor node while M signifies super/smart nodes, G represents the set of paths and the relationship between a pair of super nodes, and a pair of particles is the number of edges E in G. $E = \{(v_i, v_j)|Hop(v_i, v_j) \leq \tau\}$, where $Hop(v_i, v_j)$ is the distance between v_i and v_j. $P(v_i, v_j)$ is a path that runs from v_i to v_j in graph G. It is a sequence of edges we get when we go from v_i to v_j, where $i = j = 1, 2, \dots, N + M$. Hence, we can describe G as a set of unconventional routes $p_i(v_i, v_j)$. $e \in p_i(v_i, v_j)$, (v_N, v_{N+M}) denotes a connection between any two nodes, $E(v_i, v_j \in p_i(v_i, v_j))$ is the node disjoint between $p_i(v_i, v_j)$, (v_N, v_{N+M}) and e. Hence, we can obtain k-disjoined paths in G. We use the amount of energy consumed by a multipath, delay, and throughput to evaluate how best a multipath performs, and we use the roots on Refs. [23,24] to come up with a solution of the objective function that minimizes energy consumption and average delay, and maximizes the system throughput and network lifetime.

5.3.1 Problem Formulation

For our problem statement, we are looking to design a k-disjoined multipath for a fault-tolerant system, which uses a UAV to transmit multimedia to a super node located in a two-tiered WSN. The model is constructed in such a way that each sensor node in the network is within the transmission range of each other. This helps in minimizing the QoS parameters, such as the transmission power level and latency, while maintaining a k-disjoint multipath route. In this system, every sensor node is connected to at least one super node with k-disjoint multipath. Accordingly, a k-disjoint multipath constructed by linking a cluster of super nodes (or UAVs) with a bunch of sensors can modify their transmission range to a prime value. The transmission range of each sensor should be such that a minimum amount of energy is used while still maintaining a *k*-disjoined multipath, and all the parameters for QoS are still upheld.

5.3.2 Energy Model

For proper energy limitations, we need to consider the number of hops and the distance between two UAVs along the predefined path. The neighborhood topology mentioned in Section 5.3 is used. Each sensor node is within the transmission range of its neighbor sensor node. Given that the transmission range is equal to t_u (>0), then the neighborhood is formulated by

$$\aleph_{u,v} = \left\{ v, u \neq v \middle| \|n_u - n_v\| \leq t_u \right\} \tag{5.1}$$

It is worth recognizing that there is a chance that this might change during the dynamic network lifetime. Moreover, unless all constraints are met, there will be a division in the multipath, and neighborhood will be reconstructed. We can use the constraints to change the topology of the system that will eventually lead to solving the optimal power problem. Authors in Ref. [23] use the method of cutoff value to determine the lower and upper bound of the number of hops and transmission range. E_{elec} represents the energy deprived when using the transmitter and receiver circuitry. The energy used by the receiver to obtain a proper signal-to-noise ratio is represented by ε_{mp}. The amount of energy loss during transmission is α, and $\tau_{n(nu,nv)}$ is the transmission range. To conclude, the function used to get the minimum amount of energy used in one node to transport data of length L_p for a distant of τ is formulated by

$$min \quad \vec{Z}$$

s. t.

$$hop = \sqrt{\alpha \tau_n (n_u, n_v) \left(\frac{3\varepsilon_{mp}}{2E_{elec_{n(nu,nv)}}} \right)} \leq \tau_n (n_u, n_v) \tag{5.2}$$

$$\vec{Z} = \text{Energy}_{n_{sd}} = L_p \left\{ \sum_{n_n}^{n_d} 2 \left[E_{elec_{n_{sd}}} + \varepsilon_{mp} (n_{sd})^\alpha \right] \right\} \tag{5.3}$$

The earlier energy value for the selected path can change according to the selected upper/lower bounds, E_{min}, and E_{max}, respectively. It represents the minimum and maximum constants, respectively.

5.3.3 Delay Model

In this research, we consider the delay definition that depends on the hop count, denoted as $\varphi(\xi_i, \xi_j)$. φ represents the delay between two nodes, and its definition is determined by the ideal number of hops. Given the optimal number of hops in Eq. (5.3), which represents the minimum delay between two nodes, we can formulate and optimize the route selection while considering delay and network resource constraints. This optimization problem shall consider both source and intermediate nodes periodically in immediate neighborhood. Additionally, if one sensor node gratifies one QoS, the problem converges, and all QoS requirements will be achieved. Consequently, the end-to-end delay for a given path P between ξ_{Source} and ξ_{Sink} is described as

$$\varphi_{SourcesSink}\left(L_p\right) = min\left\{\sum_{\xi_i}\varphi\left(\xi_i,\xi_j\right)\right\}, \tag{5.4}$$

where $\varphi_{SourcesSink}$ denotes the minimum delay that we can achieve when we send data through paths between ξ_{Source} and ξ_{Sink}. This time consists of the time for transmission, retransmission, staying idle, queuing, propagation, and processing. And thus, considering

$$\sum_{v=1}\varphi\left(\xi_i,\xi_j\right) \le X_v\Delta_\varphi \tag{5.5}$$

the average delay per sensor node is equal to ξ. Assuming the hop count on a path between ξ_{Source} and ξ_{Sink} is given by η_{ij} and the delay along this path is L_e^φ. The hop delay constraint can be signified by $L_e^\varphi = \dfrac{\Delta_\varphi - \varphi^e}{\eta_{ij}}$. Accordingly, we can rewrite the constraint in Eq. (5.5) as

$$\sum_{u=1}\varphi\left(\xi_i,\xi_j\right) \le X_u L_e^\varphi \tag{5.6}$$

5.3.4 Throughput Model

According to Ref. [24] a definition of throughput can be used to represent the number of data packets successfully transmitted. This can help in calculating the optimal hop count while maximizing the network throughput. This throughput (Th) is computed via the following:

$$Th = \left(\frac{L_e^\varphi}{\xi}\right) * TR \tag{5.7}$$

where TR is the transmission rate.

5.4 PSO in IIoT

In view of the technical solutions for IIoT systems, it is necessary to describe the components and their interrelationship, namely the roles of machines, how they generate, exchange, and consume data to fulfill the interactive operations. From a swarm system perspective, industrial machines/sensors can be considered as particles with unique associated characteristics/events.

Every sensor/particle is allocated to a k-disjoint multipath according to the sensor transmission level and required separation distance per hop, as alluded in Eq. (5.3). This is performed while considering numerous attributes of swarm particles. In this approach, sensor nodes have the capacity to enhance agreeable learning conduct by trading path-related messages with their neighbors. After trading/exchanging these messages, every node/particle figures the disjoint ways and expands the neighborhood set, as alluded in Eq. (5.3). As indicated by Eqs. (5.2) and (5.4), another potential set of paths will be formed and prioritized, and thus, hops per the k-disjoint multipath will be adaptively changed according to the particle speed $v_{(i,j)}$ that is refreshed after each iteration to fulfill the desired QoS requirements.

Given that a k-disjoined multipath can have m descriptive QoS attributes, the position and velocity of the particle v are given by an m-dimensional vector $|V| = \{v_1, v_2, ..., v_N, v_{N+1}, ..., v_{N+M}\}$. The proposed swarm algorithm in this chapter contains p_{best} and g_{best}, which are the personal and global best positions, respectively. By solving Eqs. (5.2) and (5.4) in terms of the average consumed energy and average delay, we find the nodes that connect the entire searching space in every iteration. Trading control messages between the nodes can further trigger them, which are then defined as extreme and global extreme values. Consequently, this leads to the extreme value within the feasible search space towards which we progress upon after every iteration. Hence, the nodes tend to diverge are excluded. The personal best position of the swarm is brought about by the dissemination of good objective functions as denoted in Eqs. (5.2) and (5.4). These equations are concerned with the information exchange, while satisfying the constraints that are used to get the velocity and then find an ideal multipath route as stated earlier. In every path, the personal best position of a particle $v_{(i,j)}$ is given by $p_{best,v(i,j)} = (p_{(best,v1)}, (best,v2), ..., (best,vN), (best,vN+M)})$; similarly, the global best position of a particle is given by $g_{best,v(i,j)} = (g_{(best,v1)}, (best,v2), ..., (best,vN), (best,vN+M)})$. The extent to which the $p_{best,v(i,j)}$ affects the equation is given by the coefficient of constraints φ_1, similarly, the effects of the global are denoted by the coefficient of constraints φ_2. The velocity of the updated function drives what we call the CPS optimization, where $\bar{Z}$ represents the distribution of the objective function, found after satisfying the constraints mentioned earlier. x is the constriction coefficient that assists in balancing global and local probes. It is represented as

$$x = \frac{2}{\varnothing + \sqrt{\varnothing^2 - 4\varnothing}}, \text{ where } \varnothing = \varnothing_1 + \varnothing_2 > 1.$$ Eq. (5.6) is defined as the velocity

update function regarded as the momentum function, which gives the particle's/ node's present direction. Eq. (5.6) is called the social component, and it has the ability of being drawn towards the best solution as assessed by the neighbors. Eq. (5.6) represents the cognitive module, with the ability of being drawn towards earlier results which symbolize the node behavior. The only difference between CPS and FMPS optimization algorithm is the function used to update the particle velocity. This means that we not only take into account the best position of the node but also all its neighbors.

We can ignore some of the node fault-tolerance messages that can lead to trapping in local optimal solutions, by eliminating the exchanged information about the personal best $p_{best,v(i,j)}$. Consequently, this can lead to the increase of node's ability to learn from the experience of other nodes. Hence, the performance of the algorithm highly rests on the influence of nodes while satisfying the objective function. Algorithm 1 denotes the pseudocode of the proposed CPS algorithm. It finds the p_{best}'s objective function given by Eqs. (5.2) and (5.4), first in terms of the consumed energy and average delay, and then finds the least value of objective function in the p_{best}'s objective function for k-disjoined multipath. Then, it assists in avoiding velocity fit and computes the constriction value x, as shown in $\vec{v}_v$. It updates the velocity value, and finally, establishes better fault-tolerant multipath route on which the optimal nodes are chosen.

Algorithm 1: CPS

1. input: Objective functions $f(x)$;

2. $X := \{x_1,...,x_n\} := InitParticle\left(\overline{lb},\overline{ub}\right) \rightarrow \forall_p \in \{1,...,n\}: \vec{x}_n := \vec{U}\left(\overline{lb},\overline{ub}\right)$

3. $V := \{v_1,...,v_n\} := InitParticleVelocities\left(\overline{lb},\overline{ub}\right) \rightarrow \forall_p \in \{1,...,n\}:$

$$\vec{v}_n := \left(\overline{lb}-\overline{ub}\right) \otimes \vec{U}(0,1) - \frac{1}{2}\left(\overline{ub},\overline{lb}\right)$$

4. $Y := \{\vec{y}_1,...,\vec{y}_n\} := EvaluateObjectfunction(X) \rightarrow$

$$\forall_p \in \{1,...,n\}: y_n := f\left(\vec{x}_p\right)$$

5. $P := \{\vec{p}_1,...,\vec{p}_n\} := Initllocallocallyoptimal(X) \rightarrow X$

6. $P := \{p_1^f,...,p_n^f\} := InitObjeectivefunction(Y) \rightarrow Y$

7. $G := \{\vec{g}_1,...,\vec{g}_n\} := Initgloballyoptimal(P,T) \rightarrow P$

8. $G := \{g_1^f,...,g_n^f\} := Initgloballyoptimal\left(P^f,T\right) \rightarrow P^f$

9. **while** termination condition nor met do **do**

10. **for** each particle node p of n do **do**

11. $u_p = x * \left(\overrightarrow{u_p} + \overrightarrow{\Psi} \left(0, \varphi_2 \right) \otimes \left(\overrightarrow{locallyoptimal_p} - \overrightarrow{x_p} \right) \right.$

$\left. + \overrightarrow{\Psi} \left(0, \varphi_2 \right) \otimes \left(\overrightarrow{locallyoptimal_p} - \overrightarrow{x_p} \right) \right)$

12. $\overrightarrow{x_p} := \overrightarrow{x_p} + \overrightarrow{v_p}$

13. **end for**

14. $Y := EvaluateObjectivefunction \left(X, f \right)$

15. $P, P^f := Updatelocallyoptimal \ (X, Y) \rightarrow \forall_p \in \{1, ..., n\}$:

$$\overrightarrow{P_p}, p_p^f := \begin{cases} \overrightarrow{x_p}, y_i & \text{if } y_i \text{ better than } p_p^f \\ \overrightarrow{P_p}, p_p^f & \text{otherwise} \end{cases}$$

16. $G, G^f := Updategloballyoptimal \left(P, P^f, T \right) \rightarrow \forall_p \in \{1, ..., n\}$:

$$\overrightarrow{g_p}, g_p^f := best \left(P_{T_p}, P_{T_p}^f \right), \text{ where } T_p \text{ are the neighbors of } p$$

17. **End while**

5.5 Performance Ealuation

Keeping in mind the end goal to evaluate the execution of the proposed swarm method, we perform broad reenactments. We have implemented the aforementioned algorithms, such as CPMS, CPS, and FMPS optimization, using Matlab to evaluate their objective functions and visualize their outputs. We use 100 sensors and 50 UAVs dispersed uniformly in $1,000 \times 1,000 \times 100 \, m^3$ deployment space. The path loss exponent for the wireless communication model is chosen to be 2. The underlying estimation of the transmission range of sensors is set to be 100 m to assure the association among UAVs and sensor nodes while fulfilling the focus on QoS requirements. Further simulation parameters are compacted in Table 5.1.

5.5.1 Simulation Results

Figure 5.1 presents the total energy consumption in the assumed topology with a maximum of 5 hops path length. We remark that the aggregate energy utilization in the k-disjoint multipath created by the proposed PMSO approach is superior to CPS. Since settling the objective function used by CPS experiences issues in finding k-disjoint multipath after recouping from operational failures in the immense search space. Furthermore, since this outcome is being not able to substitute the arranged multipath with some other options, more energy usage is experienced.

Another critical comment that is identified by traded messages for adaptation to failure between the super nodes (UAVs) and sensors is that CPS performs essentially

Table 5.1   Assumed Parameters

Parameter	Value
Message payload	64 bytes
Data length p	2,000 bits
Transmission range	12.00 m
Tx data rate	250 kbps
E_{elec}	50 nJ/bit
Total number of UAVs	50 sensor nodes
ε_{mp}	0.0013 pJ/bit.m^2
Topology structure	Square (1,000 m × 1,000 m)
ε_{fs}	10 pJ/bit.m^2

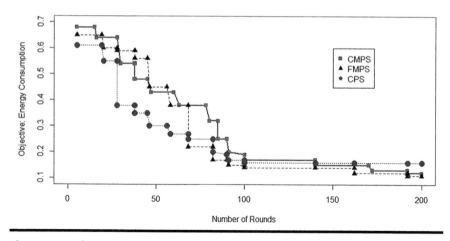

Figure 5.1   The swarm optimization routing in terms of energy consumption.

more terrible than FMPS and CPMS. That is on account that CPS necessitates fundamentally more control packets to trade among the neighbors. In this way, CPS needs to discover k-disjoint multipath in its nearby neighborhood, though the FMPS and CPMS can straightforwardly scan for routes utilizing the less control packets among the nearby hops. Subsequently, the k-disjoint multipath for FMPS and CPMS can bring down the aggregate energy usage when contrasted with the CPS approach.

Figure 5.2 demonstrates the average deferral of the chose ideal k-disjoint multipath from sender to the destination. We can watch that the assessed approaches, FMPS and CPMS, have exhibited a lower delay for every hop contrasted with CPS.

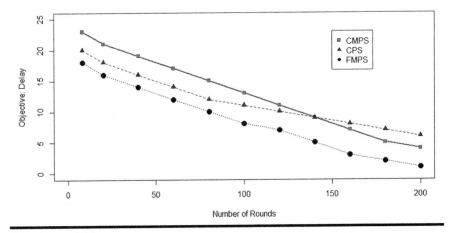

Figure 5.2   The swarm optimization routing versions vs. average delay.

This can be come back to the determination and upkeep of *k*-disjoint multipath for adaptation to failure, which can fulfill the hop necessities by choosing the following bounce in the area of every hop. Therefore, it necessitates fundamentally less control messages for adaptation to noncritical failure contrasted with CPS for choosing and keeping up 1-hop neighborhood. Hence, we can say that both CPMS and FMPS are more practical than CPS.

Throughput might vary because of high bit error rate (BER) and other surrounding changing conditions in outdoors. Accordingly, we show in Figure 5.3 the impact of tackling the objective function alluded in Eq. (5.4). We notice that, while expanding the ideal number of hops and trying to minimize the average delay, throughput degrades essentially. This is a normal influence for the assumed limited latency under the previously mentioned requirements while reducing the number

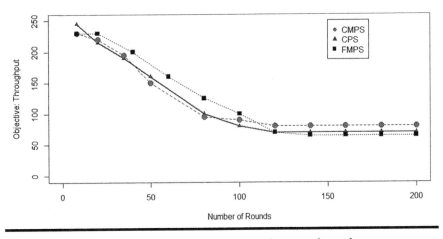

Figure 5.3   The swarm optimization routing versions vs. throughput.

of traded control messages. Therefore, the most practical hops on the route can be acquired effectively. Moreover, we presumed that the experienced algorithm behavior relies upon the IIoT network topology. Despite the fact that CPS accomplishes completely connected topology, it has shown an exceptional degraded performance in contrast to other approaches. This is due to an irregular conduct from every node while determining and limiting the multipath route hop count. At the same time, this conduct could bolster ideal execution in FMPS and CPMS with completely associated topologies.

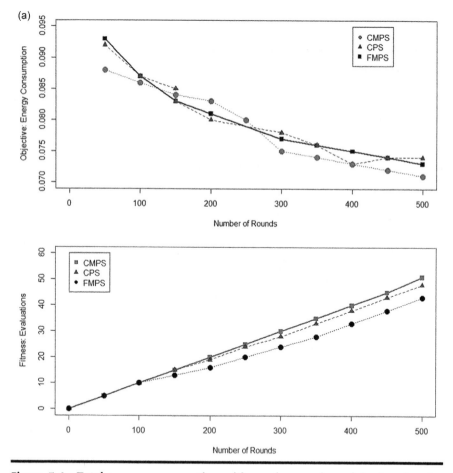

Figure 5.4 **Total energy consumption with varying number of deployed nodes with different transmission ranges. (a) Energy consumption for 30 deployed sensors. (b) Energy consumption of 40 deployed nodes. (c) Energy consumption for 50 deployed nodes.**

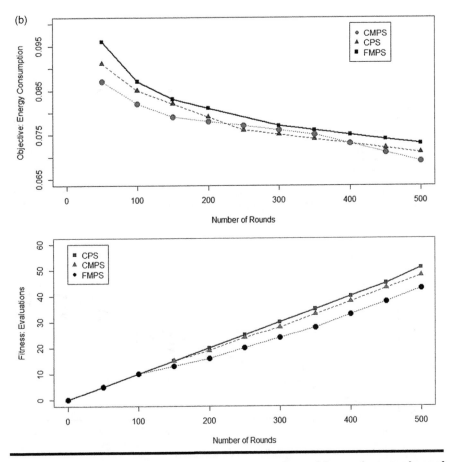

Figure 5.4 (*Continued*) Total energy consumption with varying number of deployed nodes with different transmission ranges. (a) Energy consumption for 30 deployed sensors. (b) Energy consumption of 40 deployed nodes. (c) Energy consumption for 50 deployed nodes.

We additionally explore the targeted performance when led with changing sensors/machines counts in the considered IIoT network topology. Results that have been created for every approach while expanding the quantity of sensor counts are shown in Figures 5.4–5.6. It is worth pointing out that, for multiobjective functions, the examined algorithms show better results at the beginning. Besides, the low connectivity degree (i.e., $k = 4$) could cause the algorithm to make mistakes in generating optimal approximations for the objective function. However, the behavior of particles while experiencing lower sensor counts can be utilized in estimating the objective function robustness, which could cause the algorithm to move towards more favorable regions in the feasible search space. It

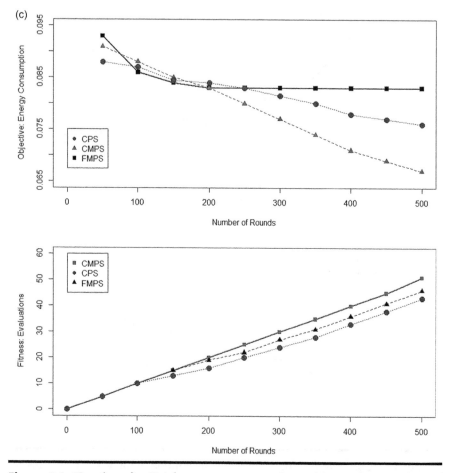

Figure 5.4 (*Continued*) **Total energy consumption with varying number of deployed nodes with different transmission ranges. (a) Energy consumption for 30 deployed sensors. (b) Energy consumption of 40 deployed nodes. (c) Energy consumption for 50 deployed nodes.**

is observed that the performance for each approach is not good at the beginning and performs better towards the end while 40 sensors or more coexist as depicted in Figures 5.4b, 5.5b, and 5.6b as well as for the 50 nodes shown in Figures 5.4c, 5.5c, and 5.6c.

According to these figures, FMPS and CPMS algorithms achieve the best performance when 30 coexisting nodes are assumed. Meanwhile, CPS performance is shown in Figures 5.4a, 5.5a, and 5.6a in terms of energy consumption, delay, and throughput, respectively, and has the worst performance when the number of nodes deployed is more than 30 at the beginning and becomes better

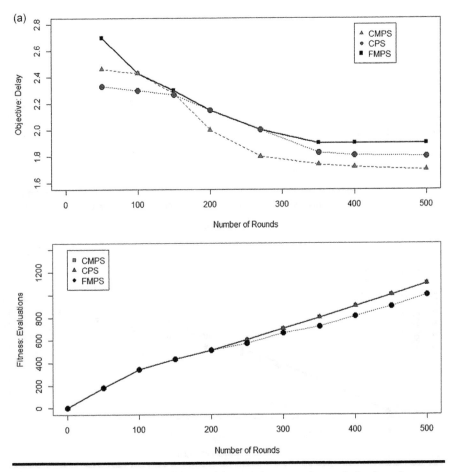

Figure 5.5 Total average delay with varying number of deployed sensor nodes with different transmission ranges. (a) Average delay for 30 deployed nodes. (b) Average delay for 40 deployed nodes. (c) Average delay for 50 deployed nodes.

over time. This is because the CPS can construct and select optimal paths from unfavorable area in the search space. Specifically, the low number of generation of paths for the objective functions can be a reason why the convergence of CPS is a little bit off from the global optimal solution for 30-node deployment than other algorithms.

Figure 5.7 compares the lifetimes of different counts of partitioning nodes, where a partitioning node is a node that can cause isolation/separation for a set of nodes in the network. In this figure, we consider a single data source (or UAV)

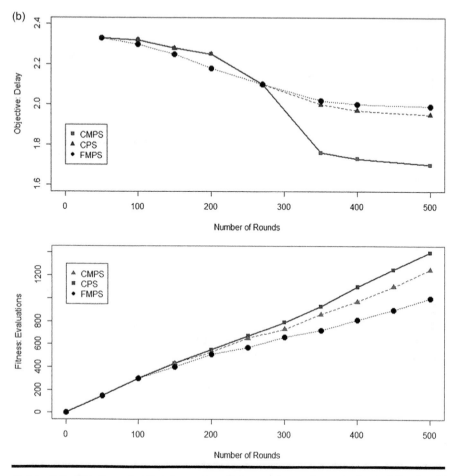

Figure 5.5 (*Continued*) **Total average delay with varying number of deployed sensor nodes with different transmission ranges. (a) Average delay for 30 deployed nodes. (b) Average delay for 40 deployed nodes. (c) Average delay for 50 deployed nodes.**

and the network lifetime definition in Ref. [24]. Accordingly, lifetime should be proportional to the ratio of total deployed nodes' count N.

Figure 5.8 illustrates the network lifetime for multiple data sources (or UAVs) and same setups in Figure 5.7. The only difference is that continuous multimedia traffic is transmitted by multiple sources to the partitioning nodes. This assumption makes the performance of proposed swarm-based algorithm easier to assess. Obviously, the network lifetime must be longer than that in the single-source scenario. This is because the network lifetime is relatively proportional to the ratio

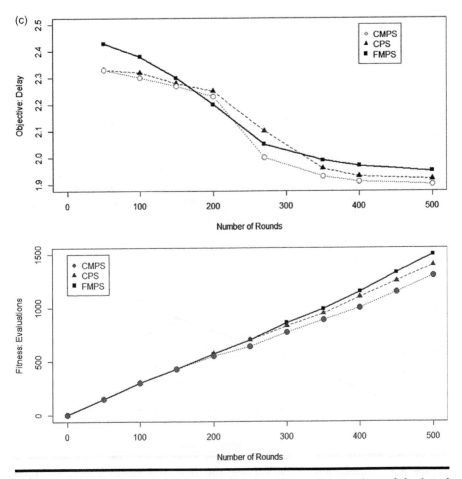

Figure 5.5 (*Continued*) Total average delay with varying number of deployed sensor nodes with different transmission ranges. (a) Average delay for 30 deployed nodes. (b) Average delay for 40 deployed nodes. (c) Average delay for 50 deployed nodes.

of the partitioning nodes to UAVs' count per region in the network. It is worth pointing out here that, with the same count of partitioning nodes, the network lifetime decreases when more than one source (UAV) is transmitting, as depicted in Figure 5.8. When more UAVs are involved in covering a region, more energy is consumed by the network per time unit. Therefore, the lifetime is expected to decrease when the number of sources increases. Similarly, in the single-source scenario (see Figure 5.7), the lifetime decreases when the partitioning node count increases.

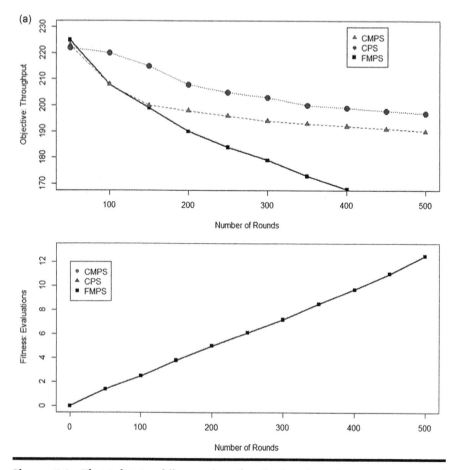

Figure 5.6 Throughput while varying the deployed nodes count. (a) Total throughput for 30 deployed nodes. (b) Total throughput for 40 deployed nodes. (c) Total throughput for 50 deployed nodes.

5.6 Conclusion

IIoT is emerging as a dominant communication paradigm nowadays to satisfy the industrial revolution worldwide. In this research, we offer a bioinspired swarm algorithm that constructs, recovers, and finds k-disjoint multipath routes in a network of machines (or UAVs). Two position information, namely the personal best position and the global position, are considered in the form of velocity update to enhance the performance of IIoT network. T validate this algorithm, we assessed the multiple objective functions that consider throughput, average energy consumption, and average end-to-end delay. Our results show that using the characteristics

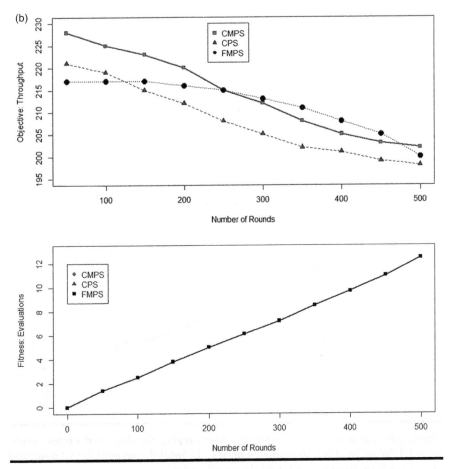

Figure 5.6 (*Continued*) Throughput while varying the deployed nodes count. (a) Total throughput for 30 deployed nodes. (b) Total throughput for 40 deployed nodes. (c) Total throughput for 50 deployed nodes.

of all personal-best information is a valid strategy for the purpose of improving CPMS performance. Moreover, the proposed algorithm has also been compared with similar algorithms, which optimize the energy consumption and average delay on the explored paths towards the destination. For the future, we see a great potential and need to study various aspects of 5G/IoT integration with the existing sensor network architectures in different levels for more successful industrial applications. The popularity of 5G, the problem of slicing the internet traffic, and the fact that a significant slice is expected to be reserved for sensory applications encourages further attempts in this domain.

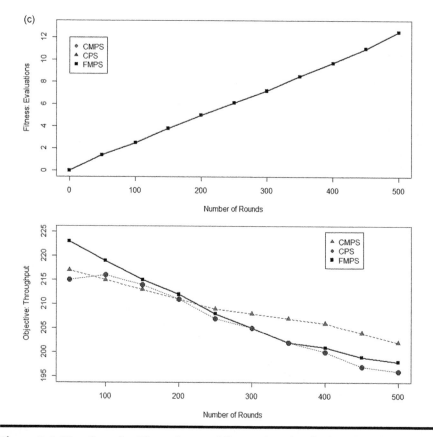

Figure 5.6 (*Continued*)   **Throughput while varying the deployed nodes count. (a) Total throughput for 30 deployed nodes. (b) Total throughput for 40 deployed nodes. (c) Total throughput for 50 deployed nodes.**

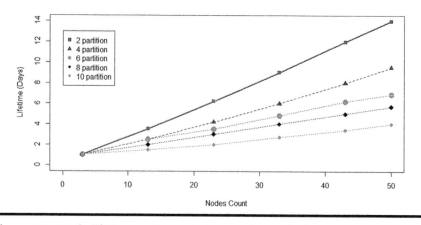

Figure 5.7   **Node lifetime vs. the nodes' count for single data source case.**

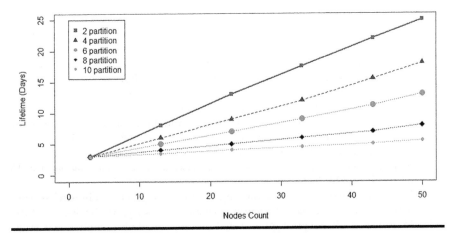

Figure 5.8 Node lifetime vs. the nodes' count for multiple data sources case.

References

1. GPP TE version 15, [On line]. www.3gpp.org/release-15. [Accessed: 25-Apr-2018].
2. F. Al-Turjman, 5G-enabled devices and smart-spaces in social-IoT: An overview, Elsevier Future Generation Computer Systems, vol. 92, no. 1, 732–744, 2019.
3. F. Al-Turjman, Fog-based caching in software-defined information-centric networks, Elsevier Computers & Electrical Engineering Journal, vol. 69, no. 1, 54–67, 2018.
4. M. Agiwal, A. Roy, and N. Saxena. Next generation 5G wireless networks: A comprehensive survey, *IEEE Communications Surveys & Tutorials,* vol. 18, no. 3, 1617–1655, 2016.
5. What is 5G? [On line]. www.surrey.ac.uk/5gic. [Accessed 25-April-2018].
6. S. Elisa, S. D. Pascoli, and G. Iannaccone, Low-power wearable ECG monitoring system for multiple-patient remote monitoring, *IEEE Sensors Journal,* vol. 16, no. 13, 5452–5462, 2016.
7. V. Petrov, et al., When IoT keeps people in the loop: A path towards a new global utility. arXiv preprint arXiv:1703.00541, 2017.
8. M. Z. Hasan, and F. Al-Turjman, Optimizing multipath routing with guaranteed fault tolerance in internet of things, *IEEE Sensors Journal,* vol. 17, no. 19, 6463–6473, 2017.
9. S. Jiang, Z. Zhao, S. Mou, Z. Wu, and Y. Luo, Linear decision fusion under the control of constrained PSO for WSNs, *International Journal of Distributed Sensor Networks,* vol. 8, no. 1, 871596, 2012.
10. F. Al-Turjman, H. Hassanein, and M. Ibnkahla, Towards prolonged lifetime for deployed WSNs in outdoor environment monitoring, *Elsevier Ad Hoc Networks Journal,* vol. 24, no. A, 172–185, January 2015.
11. J. K. Vis, Particle swarm optimizer for finding robust optima, Leiden. www.liacs.nl/assets/Bachelorscripties/2009-12JonathanVis.pdf, January 15, 2015.
12. W. H. Lim, and N. A. Mat Isa, Particle swarm optimization with adaptive time-varying topology connectivity, *Applied Soft Computing,* vol. 24, 623–642, 2014.

13. Q. Chi, H. Yan, C. Zhang, Z. Pang, and L. Xu, A reconfigurable smart sensor interface for industrial WSN in IoT environment, *IEEE Transactions on Industrial Informatics*, vol. 10, no. 2, 1417–1425, 2014.
14. B. Karschnia, Industrial Internet of Things (IIoT) benefits, examples|Control Engineering, Controleng.com, 2017. [Online]. Available: www.controleng.com/single-article/industrial-internet-of-things-iiot-benefits-examples/a2fdb5aced1d77 9991d91ec3066cff40.html. [Accessed 31-Aug-2017].
15. F. Al-Turjman, Price-based data delivery framework for dynamic and pervasive IoT, *Elsevier Pervasive and Mobile Computing Journal*, vol. 42, 299–316, 2017.
16. Y. Al-Nidawi, H. Yahya, and A. H. Kemp. Tackling mobility in low latency deterministic multihop IEEE 802.15.4e sensor network, *IEEE Sensors Journal*, vol. 16, no. 5, 1412–1427, 2016.
17. M. Dhir, A survey on fault tolerant multipath routing protocols in wireless sensor networks, *Global Journal of Computer Science and Technology*, vol. 15, issue 3, 2016.
18. M. Adnan, M. Razzaque, I. Ahmed, and I. Isnin, Bio-mimic optimization strategies in wireless sensor networks: A survey, *Sensors*. vol. 14, issue 1, 299–345, 2014.
19. F. Al-Turjman, Information-centric sensor networks for cognitive IoT: An overview, *Annals of Telecommunications*, vol. 72, no. 1, 3–18, 2017.
20. H. -L. Shieh, C. -C. Kuo, and C. -M. Chiang, Modified particle swarm optimization algorithm with simulated annealing behavior and its numerical verification, *Applied Mathematics and Computation*, vol. 218, no. 8, 4365–4383, 2011.
21. Y. Zhou, X. Wang, T. Wang, B. Liu, and W. Sun, Fault-tolerant multi- path routing protocol for WSN based on HEED, *International Journal of Sensor Networks*, vol. 20, no. 1, 37–45, 2016.
22. M. Z. Hasan, and F. Al-Turjman, SWARM-based data delivery in Social Internet of Things, *Elsevier Future Generation Computer Systems*, vol. 92, no. 1, 821–836, 2019.
23. M. Hasan, F. Al-Turjman, and H. Al-Rizzo, Optimized multi-constrained quality-of-service multipath routing approach for multimedia sensor networks, *IEEE Sensors Journal*, vol. 17, issue 7, 2298–2309, 2017.
24. F. Al-Turjman, QoS–aware data delivery framework for safety-inspired multimedia in integrated vehicular-IoT, *Elsevier Computer Communications Journal*, vol. 121, 33–43, 2018.
25. F. Al-Turjman, Cognitive routing protocol for disaster-inspired Internet of Things, *Elsevier Future Generation Computer Systems*, 2017. doi:10.1016/j.future.2017.03.014.
26. G. Singh, and F. Al-Turjman, Learning data delivery paths in QoI-aware information-centric sensor networks, *IEEE Internet of Things Journal*, vol. 3, no. 4, 572–580, 2016.
27. M. Z. Hasan, H. Al-Rizzo, and F. Al-Turjman, A survey on multipath routing protocols for QoS assurances in real-time multimedia wireless sensor networks, *IEEE Communications Surveys and Tutorials*, vol. 19, no. 3, 1424–1456, 2017.

Chapter 6

Drones Navigation in Mission Critical Applications

Fadi Al-Turjman

Antalya Biim University

6.1 Introduction

Safety and mission critical (MC) services are often the most demanding in our community [1,2]. Civilized cities are suffering from various types of disasters at risk regions. A number of events with high probability of casualties, including major earthquakes and snow avalanches, are also serious problems, especially on highways. In total, there were 1,325 fatalities caused by the avalanches between 1951 and 2007. Furthermore, over 1.2 million people were killed in road traffic accidents around the world in 2018 and another 50 million may be left injured by crashes annually. Moreover, it was reported that road traffic accidents could outstrip stroke as one of the main causes of preventable deaths by 2020 [3].

Over the last decade, the evolution of the sensor network paradigm has introduced a well-established area of MC applications. Wireless sensor networks (WSNs) provide an interesting field of research in different MC domains, ranging from the hardware design and resource awareness for effective wireless communication to deployment and positioning [4]. In environmental monitoring, sensor networks are used to observe various atmospheric parameters or to track the movement of species at risk in outdoor environments. Other applications of WSNs include health care and smart environment sensing. First-generation sensor nodes have limited abilities, such as providing scalar measurements, for parameters such as temperature, humidity, movement, lightning conditions, pressure, and noise level [5]. The primary function of these sensors is to collect information over extended periods of time and send it to a particular base station, and occasionally, raise an alarm in case of an unusual occurrence. On the other hand, second-generation sensors have gone to a whole new level. They have better sensing power, such as the use of cameras and microphones. Consequently, this led to the development of wireless multimedia sensor networks (WMSNs). Later on, it was proven that this kind of data require increased power and storage capacity as well as a wider communication bandwidth [6,7]. Accordingly, a new type of these networks, called the flying ad hoc sensor networks (FANETs), has emerged. In FANETs, the sensing and communication capabilities of a WSN have been combined with large-scale unmanned aerial vehicles (UAVs) as enabling technologies. UAVs are flying robots that can be remotely supervised through a software-embedded system working in conjunction with built-in global positioning system (GPS) modules for positioning (Figure 6.1).

However, considering the limited GPS accuracy and the line-of-sight requirement as well astheir restricted resources, such as energy and communication range, makes their accurate positioning a challenge. In Ref. [8] authors detailed these challenges and focused on optimal UAV placement. An integer linear program (ILP), taking into account limited resources in UAVs, has been proposed. It involves the available communication bandwidth and the energy used. However, the entire process of positioning is done using relative coordinates. This was significantly degrading the proposed approach performance. Other attempts have been performed based on Kalman filter (KF). There are several inadequacies in the KF technique. Such inadequacies include (1) observability problems, (2) error modeling challenges, and poor prediction during GPS outages. To avoid these problems and to provide an integrated system that can be independent from the underlying navigation system, machine learning (ML) solution has been used to solve this problem. In general, the main objective of ML is to learn the pattern of integrated data and build the path accordingly. The objective is to use this knowledge, when

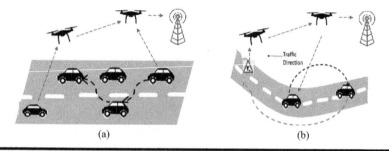

(a)   (b)

Figure 6.1 (a) Traffic overtaking on bridges, (b) sharp corners, and blind spot risks can be propagated to other autonomous vehicles via the means of UAVs flying on top, thus preventing potential collisions.

GPS signal is not available, to predict the trajectory path. A possible model of the multilayer perceptron architecture can be based on a neural network (NN) approach that computes a nonlinear function of the scalar product of the input vector and the weight vector. An alternative architecture of NN is one in which the distance between the input vector and a certain prototype vector determines the activation of a hidden NN unit.

In this chapter, we propose a two-level KF-based positioning system to integrate inertial navigation system (INS) and GPS measures and provide an intelligent learning approach. The presented system enables more accurate positioning in critical applications. The system involves training the network in real time, when the GPS signal is available, and predicting proper positioning, when GPS signal is not available.

The remainder of this chapter is organized as follows. In the next section, a literature review is presented. Section 6.3 describes system models used to implement the KF system. In Section 6.4, we discuss the methodology followed in this system. Experimental results and discussions are detailed in Section 6.5. Moreover, conclusions are provided in Section 6.6. In the following, Table 6.1 summarizes all used abbreviations in this study to further assist the reader.

Table 6.1 Table of Used Abbreviations and Definitions

Abbreviation	Definition
IoT	Internet of things
KF	Kalman filter
RFID	Radio frequency identifier
MLE	Maximum likelihood estimation
UAV	Unmanned aerial vehicle
RSSI	Received signal strength indicator
INS	Inertial navigation system
GPS	Global positioning system
MC	Mission critical
LoS	Line of sight
QoS	Quality of service
ILP	Integer linear program
ML	Machine learning

(Continued)

Table 6.1 (Continued) Table of Used Abbreviations and Definitions

Abbreviation	Definition
NN	Neural network
FANETs	Flying ad hoc networks
WSN	Wireless sensor networks
WMSN	Wireless multimedia sensor networks
MC	Mission critical
TLKAPA	Two-level KF-aided positioning approach
DR	Dead reckoning
DGPS	Differential GPS
RSU	Roadside unit
IMU	Inertial measurement unit
RS	Reference station
RF	Radio frequency

6.2 Literature Review

These days, distinctive UAV positioning strategies, for example, satellite, roadside reference points, dead reckoning (DR), and image processing are being used. Satellite-based positioning is otherwise called GPS. GPS comprises of somewhere around four satellites as hotspots for the present position computations and a GPS recipient in the UAV. As of now, this GPS innovation is a standout amongst the most generally utilized positioning approaches in UAV frameworks [7,9]. GPS utilizes two primary standards for situating and routing: the code stage and the transporter stage. In the first stage, information is accumulated by means of a coarse obtaining code to ascertain the UAV position. In the second stage, the satellite radio signal is utilized to refresh the determined position when a line of sight (LoS) is available. GPS has the capacity to accomplish exactness of 5–10 m after removing the noise data [9].

In the meantime, differential GPS (DGPS) innovation came to enhance the standard unadulterated GPS estimations. Right off the bat, standard GPS mistakes are determined explicit at surely understood position called reference station (RS). The differential revisions are then transmitted through satellites for compensation purposes in the continuous position computations [10]. However, GPS innovation can't deal with boundaries, as the RF signals are transmitted through high-recurrence microwaves and comes up short on the capacity to go through obstacles.

This outcome leads to navigation blackouts or inaccurate positioning. Another GPS disadvantage can be the electromagnetic impedance impacts brought about by different remote signs concurrence, which can significantly degrade UAV performance.

On the other hand, DR, otherwise called INS, is another UAV situating approach, where the present position is assessed dependent on recently determined positions, time distinction, and accelerated speed [10,11]. The primary sensors utilized in DR innovation are accelerometers and gyroscopes. Unlike the GPS, DR does not rely on any outside referencing signal. Its refresh time is multiple times more than the GPS every second. Moreover, the advanced shabby gyroscopes and accelerometers have much better refresh/update rates. This capacity is useful for applications where UAV tracking is required under unique conditions. Likewise, a constant reset of the UAV position is required at normal interims and well-known positions [11]. This reset is required for error control as separation and speed errors can accumulate over time. Unfortunately, DR innovation isn't prescribed for unpredicted situations, for example, the harsh climate conditions in open outdoor areas, since it can prompt severe errors in the actual position computations.

Roadside beacon is another technology used in determining the location of UAV through simple sensors placed along the UAV trajectory [12]. These sensors sense the proximity of UAV with respect to their reference positions. Microwave, radio frequency (RF) identifiers (RFIDs), infrared, or RF roadside units (RSUs) with the capability to receive and transmit data to UAV can be employed to generate the beacon signal [12–15]. This can be achieved in two methods: self-positioning or remote positioning. In self-positioning, the UAV collects signals from the beacon sensors. However, in remote positioning, the beacon unit reads the UAV through a tag or transmitter placed on its body. In this method, performance of the roadside beacon positioning is strongly dependent on the count of RSUs. The smaller the distance between RSUs, the better its positioning availability and continuity. Unluckily, this can dramatically increase the cost of such alternatives for UAV positioning systems.

Meanwhile, image processing is another positioning strategy for UAVs. It is comprehensively used in rush-hour traffic systems where the navigated region can be recognized by applying ML methods [15–18]. In any case, this system has not yet turned out to be mainstream in UAV applications, as it necessitates a significant amount of bandwidth and processing capabilities to exchange these images. Moreover, extracting features of interest from these images in a real-time manner is not an easy task. Hence, such techniques are not recommended in real-time UAV applications such as those targeted in MC scenarios.

In this chapter, we propose a two-level KF approach to overcome the positioning problems in the literature. INS and GPS position data are fused together to predict the INS positioning errors. Moreover, the proposed system predicts the UAV position during GPS outages, based on processing only INS position components. We implement and simulate the earlier proposed system while not including velocity data. This is done to avoid further computational load in real-time mode.

Moreover, a modified KF algorithm has been introduced to reduce the increase in prediction errors over time.

6.3 System Models

Basically, the implementation of navigation system in any UAV depends on three important parts [18–20]. First, the mechanization part that transforms the real measurements of the accelerometers and gyros from the UAV body frame into the local frame that is the stable earth-fixed coordinates. Second, the error model part describes and takes into consideration the errors that may affect our predicted output. Third, the filter part will help in removing the noise from the calculated positioning components. To further assist the reader, Table 6.2 summarizes our used notations in this study.

Table 6.2 Table of Used Notations

Abbreviation	Definition
A, r, and p	The three angular azimuth, pitch, and roll quantities
V^e, V^n, and V^u	The velocities in east, north, and up directions
h, φ, and λ	The three axis altitude components
Ω_{ie}^l, f^l, and g^l	These are the components for the earth rotation, the local-level frame change of orientation, and gravity
r^l, V^l, R_b^l	The local-level mechanization equations for distance, velocity and rotation.
Q	Quaternion parameters matrix
θ_x, θ_y, and θ_z	The angles between the UAV and the x, y, and z axes
x	The error state vector
δf^b and $\delta\omega^b$	The accelerometers and gyroscopes bias errors
ω_x, ω_y, and ω_z	The angular velocities of the UAV body rotation
f_x, f_y, and f_z	The x-, y-, z-axis accelerometers measurements
Ω_{ib}^b	The local-level frame change of orientation
F and x	These are the state transition matrix and error states, respectively
Gw	This is the system noise
K	Kalman gain

6.3.1 Mechanization Equations

In the following, we detail the assumed mechanization matrices and their corresponding derivatives [17]. First, we start with the rotation matrix R_b^l:

$$R_b^l = \begin{pmatrix} \cos A \cos r - \sin A & -\sin A \cos p & \cos A \sin r + \sin A \\ \sin A \cos r - \cos A & \cos A \cos p & \sin A \sin r - \cos A \\ -\cos p \sin r & \sin p & \cos p \cos r \end{pmatrix} \quad (6.1)$$

where A, r, and p are three angular azimuth, pitch, and roll quantities, respectively. As for the positioning mechanization equations, we use the following:

$$r^l = \begin{pmatrix} \varphi \\ \lambda \\ h \end{pmatrix} = \begin{pmatrix} 0 & \dfrac{1}{\cos \varphi} & 0 \\ \dfrac{1}{(R+h)\cos \varphi} & 0 & 0 \\ 0 & 0 & 1 \end{pmatrix} \begin{pmatrix} V^e \\ V^n \\ V^u \end{pmatrix} = D^{-1} V^l \quad (6.2)$$

where V^e, V^n, and V^u are the velocities in east, north, and up directions. h, φ, and λ are the three axis altitude components. As for the velocity equations,

$$V^l = R_b^l f^b - \left(2`\Omega_{ie}^l + `\Omega_{ie}^l\right) V^l + g^l, \quad (6.3)$$

where components for the earth rotation, the local-level frame change of orientation, and gravity $`\Omega_{ie}^l, f^l$, and g^l are given by

$$f^l = \begin{pmatrix} f^e \\ f^n \\ f^u \end{pmatrix} = R_b^l f^b = R_b^l \begin{pmatrix} f_x \\ f_y \\ f_z \end{pmatrix}, \text{ and } g^l = \begin{pmatrix} 0 \\ 0 \\ -g \end{pmatrix}, \quad (6.4)$$

$$`\Omega^l_{ie} = \begin{pmatrix} 0 & -\omega^e \sin\varphi & \omega^e \cos\varphi \\ \omega^e \sin\varphi & 0 & 0 \\ -\omega^e \cos\varphi & 0 & 0 \end{pmatrix} `\Omega^l_{ie}$$

$$= \begin{pmatrix} 0 & \dfrac{-V^e \tan\varphi}{N+h} & \dfrac{V^e}{N+h} \\ \dfrac{V^e \tan\varphi}{N+h} & 0 & \dfrac{V^\eta}{M+h} \\ \dfrac{-V^e}{N+h} & \dfrac{-V^\eta}{M+h} & 0 \end{pmatrix} \tag{6.5}$$

As for the altitude equations, we assume the following:

$$R^l_b = R^l_b \left(`\Omega^b_{ib} - `\Omega^b_{il} \right) \tag{6.6}$$

where

$$\omega^b_{il} = R^l_b \left[\begin{pmatrix} 0 \\ \omega^e \cos\varphi \\ \omega^e \sin\varphi \end{pmatrix} + \begin{pmatrix} \dfrac{-V^\eta}{M+h} \\ \dfrac{V^e}{N+h} \\ \dfrac{V^e \tan\varphi}{N+h} \end{pmatrix} \right]$$

$$= R^l_b \begin{pmatrix} \dfrac{-V^\eta}{M+h} \\ \dfrac{V^e}{N+h} + \omega^e \cos\varphi \\ \dfrac{V^e \tan\varphi}{N+h} + \omega^e \sin\varphi \end{pmatrix} \tag{6.7}$$

and

$$\omega^b_{ib} = \begin{pmatrix} \omega_x \\ \omega_y \\ \omega_z \end{pmatrix} \tag{6.8}$$

Therefore, the local-level mechanization equations for distance, velocity, and rotation r^l, V^l, and R_b^l, respectively, can be calculated as follows:

$$\begin{pmatrix} r^l \\ V^l \\ R_b^l \end{pmatrix} = \begin{pmatrix} D^{-1}V^l \\ R_b^l f^b - \left(2\dot{\Omega}_{ie}^l + \dot{\Omega}_{el}^l\right)V^l + g^l \\ R_b^l \left(2\dot{\Omega}_{ib}^b + \dot{\Omega}_{il}^b\right) \end{pmatrix} \tag{6.9}$$

Then, Quaternion method is used to solve these mechanization equations. Quaternion parameters can be expressed as follows:

$$Q = \begin{pmatrix} q_1 \\ q_2 \\ q_3 \\ q_4 \end{pmatrix} = \begin{pmatrix} \left(\dfrac{\theta_x}{\theta}\right)\sin(\theta/2) \\ \left(\dfrac{\theta_y}{\theta}\right)\sin(\theta/2) \\ \left(\dfrac{\theta_z}{\theta}\right)\sin(\theta/2) \\ \cos(\theta/2) \end{pmatrix} \tag{6.10}$$

where θ_x, θ_y, and θ_z, are the angles between the UAV and the x, y, and z axes. To determine the UAV body velocity change at time t_{k+1}, we use the following equations:

$$V^l(t_{k+1}) = V^l(t_k) + \frac{1}{2}\left(\Delta V^l(t_k) + \Delta V^l(t_{k+1})\right) \tag{6.11}$$

where

$$V^l = \begin{pmatrix} V^e \\ V^n \\ V^u \end{pmatrix} \tag{6.12}$$

To calculate the aforementioned altitude components,

$$h(t_{k+1}) = h(t_k) + \frac{1}{2}\left(V^u(t_{k+1}) + V^u(t_k)\right)\Delta t \tag{6.13}$$

$$\varphi(t_{k+1}) = \varphi(t_k) + \frac{1}{2}\frac{\left(V^n(t_{k+1}) + V^n(t_k)\right)}{R+h}\Delta t \tag{6.14}$$

$$\lambda(t_{k+1}) = \lambda(t_k) + \frac{1}{2} \frac{\left(V^n(t_{k+1}) + V^n(t_k)\right)}{(R+h)\cos\varphi} \Delta t \qquad (6.15)$$

Since $h = V^u$, we get the following:

$$\emptyset = \frac{V^n}{R+h}, \quad \lambda = \frac{V^e}{(R+h)\cos\varphi} \qquad (6.16)$$

6.3.2 Error Model

In this chapter, the error state vector is determined by Eq. (6.17).

$$x = \begin{pmatrix} \delta\varphi & \delta\lambda & \delta h & \delta V^e & \delta V^n & \delta V^u & \delta p & \delta r & \delta A & \delta f_x & \delta f_y & \delta f_z & \delta\omega_x & \delta\omega_y & \delta\omega_z \end{pmatrix}^T \qquad (6.17)$$

Eq. (6.17) classified these errors into coordinate errors (δr^l), velocity errors $\begin{pmatrix} \delta V^e \\ \delta V^n \\ \delta V^u \end{pmatrix}$, and altitude errors $\begin{pmatrix} \delta p \\ \delta r \\ \delta A \end{pmatrix}$. These errors are modeled in Eqs. (6.18)–(6.20) as follows.

$$\delta r^l = \begin{pmatrix} \delta\varphi \\ \delta\lambda \\ \delta h \end{pmatrix} = \begin{pmatrix} 0 & \dfrac{1}{M+h} & 0 \\[3mm] \dfrac{1}{(N+h)\cos\varphi} & 0 & 0 \\[3mm] 0 & 0 & 1 \end{pmatrix} \begin{pmatrix} \delta V^e \\ \delta V^n \\ \delta V^u \end{pmatrix} \qquad (6.18)$$

$$\begin{pmatrix} \delta V^e \\ \delta V^n \\ \delta V^u \end{pmatrix} = \begin{pmatrix} 0 & f_u & -f_\eta \\ -f_u & 0 & f_e \\ f_\eta & -f_e & 0 \end{pmatrix} \begin{pmatrix} \delta p \\ \delta r \\ \delta A \end{pmatrix} + R_b^l \begin{pmatrix} \delta f_x \\ \delta f_y \\ \delta f_z \end{pmatrix} \qquad (6.19)$$

$$\begin{pmatrix} \delta p \\ \delta r \\ \delta A \end{pmatrix} = \begin{pmatrix} 0 & \dfrac{1}{R+h} & 0 \\[3mm] \dfrac{-1}{R+h} & 0 & 0 \\[3mm] \dfrac{-\tan\varphi}{R+h} & 0 & 0 \end{pmatrix} \begin{pmatrix} \delta V^e \\ \delta V^n \\ \delta V^u \end{pmatrix} + R_b^l \begin{pmatrix} \delta\omega_x \\ \delta\omega_y \\ \delta\omega_z \end{pmatrix}. \qquad (6.20)$$

In Eqs. (6.21)–(6.22), the accelerometers (δf^b) and gyroscopes ($\delta \omega^b$) bias errors are modeled.

$$\delta f^b = \begin{pmatrix} \delta f_x \\ \delta f_y \\ \delta f_z \end{pmatrix}$$

$$= \begin{pmatrix} -\alpha_x & 0 & 0 \\ 0 & -\alpha_x & 0 \\ 0 & 0 & -\alpha_x \end{pmatrix} \begin{pmatrix} \delta f_x \\ \delta f_y \\ \delta f_z \end{pmatrix} + \begin{pmatrix} \sqrt{2\alpha_x \sigma_x^2} \\ \sqrt{2\alpha_y \sigma_y^2} \\ \sqrt{2\alpha_z \sigma_z^2} \end{pmatrix} \omega(t) \qquad (6.21)$$

$$\delta \omega^b = \begin{pmatrix} \delta \omega_x \\ \delta \omega_y \\ \delta \omega_z \end{pmatrix}$$

$$= \begin{pmatrix} -\beta_x & 0 & 0 \\ 0 & -\beta_x & 0 \\ 0 & 0 & -\beta_x \end{pmatrix} \begin{pmatrix} \delta \omega_x \\ \delta \omega_y \\ \delta \omega_z \end{pmatrix} + \begin{pmatrix} \sqrt{2\beta_x \sigma_x^2} \\ \sqrt{2\beta_y \sigma_y^2} \\ \sqrt{2\beta_z \sigma_z^2} \end{pmatrix} \omega(t) \qquad (6.22)$$

where ω_x, ω_y, and ω_z are the angular velocities of the UAV body rotation that are determined by gyroscopes after compensating for earth rotation and local-level frame change of orientation $\left(\Omega_{ib}^b\right)$. And f_x, f_y, and f_z are the x-, y-,and z-axis accelerometers measurements, respectively. These error models are required for the analysis and estimation of different error sources associated with the localization system assumed in this chapter, as described in the following sections.

6.3.3 Filtering Model

KF is a key component in the proposed approach, as it eliminates undesired noise in sensors' (accelerometers and gyroscopes) readings and/or measurements. It uses an independent source of readings, which is the INS system. In Figure 6.2 we overview the functionality of KF using a block diagram.

Accordingly, the different INS errors can be described using the following state transition matrix:

$$\dot{x} = Fx + Gw \qquad (6.23)$$

where F and x are the state transition matrix and the error states, respectively, and Gw is the system noise.

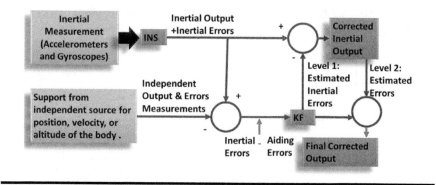

Figure 6.2   Illustrative block diagram for the used KF.

6.4 Two-Level KF-Aided Positioning Approach for UAVs

Describing the motion of an UAV on the earth's surface or close to it has become a major issue nowadays. Since the motion in general can be described by six parameters (three position and three orientation parameters), the UAV navigation system requires the capability to sense six independent quantities from which these parameters can be derived. These quantities are measured using two types of sensors (accelerometers and gyroscopes). For navigation in 3D, three accelerometers and three gyroscopes are required. The gyroscopes are used for monitoring angular motion in three directions (north, east, and up direction). The accelerometers are used for monitoring linear motion in the same three directions. The transformation is done using the aforementioned mechanization equations in Section 6.3. The accelerometers are used for linear motion monitoring in the same orthogonal x, y, and z-axes directions. The inertial measurement unit (IMU) is attached to the UAV body. Since we need the acceleration to be in the same frame as the coordinate system, transformation of it from the UAV body frame (b-frame) to the stable earth-fixed frame (l-frame) is required. That transformation is done using mechanization equations. The most notable navigation systems are the INS and GPS. INS and GPS represent two basic forms of navigation. As aforementioned, INS does not rely usually on reference points. It depends only on the initial UAV velocity and position information. On the other hand, GPS and radio navigation in general depends on a well-known reference point, such as a satellite in the earth orbit. Integrating both INS and GPS have several advantages. One of the most important advantages is that they complement the errors of each other. Therefore, in this chapter, we are testing the results of integrating the INS and GPS systems using two-level KF in comparison with the unaided INS and GPS. In the first level, KF is used to estimate and eliminate INS system errors. It compromises two modes of operation: (1) the training mode and (2) the prediction mode. Our reference in this level shall be

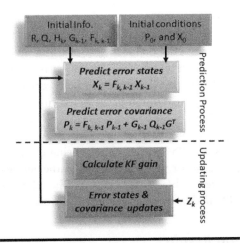

Figure 6.3   Illustrative block diagram for the proposed TLKAPA method.

the output of an accurate INS. In the second level, it is used to detect and eliminate the integrated INS/GPS errors for more accurate location information for the flying UAV. It represents the updating mode in our proposed two-level KF-aided positioning approach (TLKAPA) method, and it functions as long as the GPS signal is available. And hence, the overall workflow of the proposed approach can be summarized as depicted in Figure 6.3.

In the following section, we assess the performance of the proposed TLKAPA approach via extensive simulation results that relies on real GPS data in practice.

6.5 Performance Evaluation

To simulate the GPS outage and to predict it properly, a KF is used as described earlier. GPS signal is simulated to be 0 in case of an outage, and then KF is used to predict the error in two levels; first, in INS signals, and second, in integrated GPS/INS errors. The performance of the KF-based module was examined using four simulated GPS outages. The locations of these GPS outages were chosen to examine the performance of KF module with respect to different UAV node dynamics.

6.5.1 Setup and Simulation Environment

Using MATLAB®, we simulate GPS outage by assigning 0 to the GPS data signal and simulate prediction of GPS signal using MATLAB neural network toolbox (NEWRB) Function. The simulation is based on predicting the error between GPS and INS signals. This predicted error is then used to calculate GPS signal, hence arriving at proper positioning. The presented system is tested using real measurements from inertial Xbow sensors and GPS mounted on a UAV. To evaluate the predicted path, the original plot (where GPS signal is available) is plotted against the predicted path for the same trajectory. All obtained figures are run under the Gauss-Markov error model and INS/GPS combined reference solution as an aiding source. Gaussian functions in this simulation are all assumed to be 0 centered with 1° of standard deviation. This is the default definition of NEWRB function in MATLAB. These defaults work fine for the purpose of this simulation. The learning window size is assigned to be 30 in the input for NEWRB in MATLAB. This is the best size that captures the dynamics of GPS signals, without being overloaded in computation. While the training window size was 30 s (1 epoch/s), the outage duration was 30 s (1 epoch/s). Altitude was ignored (assumed to be 0). The reference points were actual datasets from the GPS module for the same period of outage simulation.

6.5.2 Simulation Results

In this section we test the effect of different factors that may affect the KF and unaided outputs of the 3D simulator. So we try to see the effect of accelerometers and gyros bias increment and decrement, and GPS outage effects. We discuss the effect of bias on the different error states and measured components. By running the 3D simulator, we were able to show the Schuler effects of bias, as you can see in Figure 6.4.

In Figure 6.4a, you can see the sinusoidal effect of the bias. Also you can see the decrease of the difference between the output of the KF and the output of our

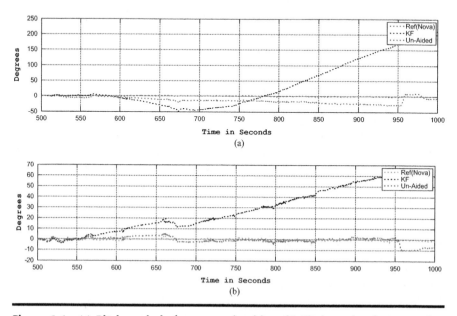

Figure 6.4 (a) **Pitch angle before removing bias.** (b) **Pitch angle after removing bias.**

reference, where it is 15° at $t = 700$ s without bias (in Figure 6.4b) and 50° at the same time with bias (in Figure 6.4a). That bias effect on the deference is due to the strong coupling between pitch error δp and the north velocity error δV^n, as derived in Section 6.3 equations. Similarly, we can show the Schuler effect of the bias on the azimuth and roll angles. Also, in Figure 6.5, we can see the effect of bias on the velocity in the east and north directions, especially because of the strong coupling between the pitch and roll errors and the north velocity error δV^n and the east velocity error δV^e. So, it's clear that there is no strong bias effect on the vertical velocity, while the velocity in east and north are strongly affected. As for the positioning components, latitude, longitude, and altitude, they are also affected by the bias removal, as shown in Figure 6.6. The least affected positioning component by the bias is the altitude. Moreover, it's clear from Figure 6.7 that the latitude error is larger than the longitude error, because it's directly related to the north velocity error while the longitude is not directly related to the east velocity error as shown in $\frac{1}{(R+h)} \delta V^n = \delta \varphi$ and $\frac{1}{(R+h)\cos\varphi} \delta V^e = \delta \lambda$. Meanwhile, by inserting four GPS outages, we got the results in Figure 6.8. It's clear that the KF output in the pitch, roll, and azimuth start diverging earlier because of the outages of GPS that makes the errors accumulated together to form a huge difference between the KF output and the reference (in red color).

6.6 Conclusions

In this chapter, the real-time INS and GPS integration in UAV navigation utilizing a two-level KF has been simulated and examined. This approach is based on predicting the INS position error and continuously removing it from its corresponding INS position, in addition to a second level of applied KF for the overall integrated INS/GPS errors. Results showed the ability of the KF-based module to reduce the INS position error and prevent its growth even in the long term. We saw in the cases, error is stabilized and takes a pattern in time. In addition, the proposed KF module was able to accurately predict the INS position errors during GPS outages. This was the objective of the system, as introduced early in this chapter. Results point out to success in achieving these objectives, of being able to predict, in real time, the proper position of UAV, based on learning the dynamics of the most recent GPS signal, with good degree of accuracy. The simulation proved that the integration technique is a reliable, robust, and self-adaptive system that requires no prior knowledge of the navigation system utilized. We conclude from this study that several factors can affect the performance of the KF output (prediction). But still we can achieve a high-performance navigation system by integrating INS with GPS s. It outperforms the unaided INS or GPS system even if we used Xbow sensors that have lower accuracy than Novatel.

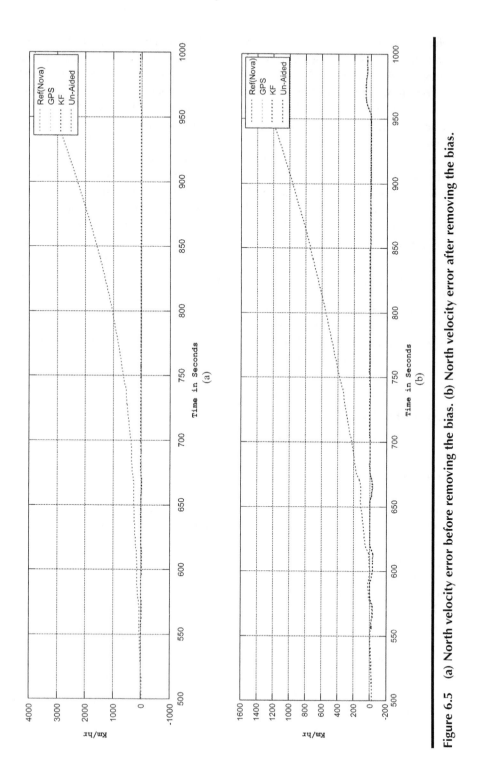

Figure 6.5   (a) North velocity error before removing the bias. (b) North velocity error after removing the bias.

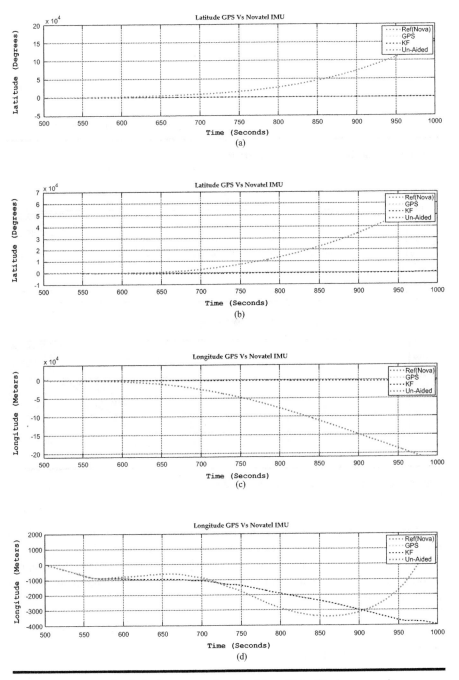

Figure 6.6 (a) Latitude before removing the bias. (b) Latitude error after removing the bias. (c) Longitude error before removing the bias. (d) Longitude error after removing the bias.

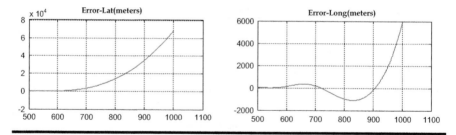

Figure 6.7   Latitude vs. longitude error.

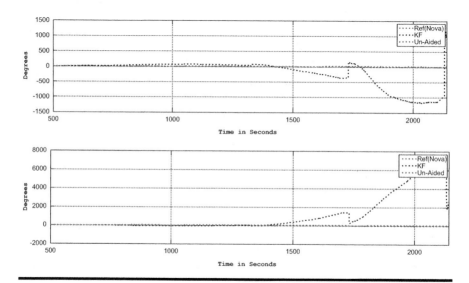

Figure 6.8   Effect of the four GPS outages.

A future research could focus on what are the reasons where dynamics of GPS caused that error in the second case, above predicting the path. Moreover, it could also investigate the third dimension (altitude) that was not covered in this chapter. Another area of improvement in future research is further reducing the increase in prediction error over time.

Acknowledgements

We thankfully acknowledge all the help and support of Prof. A. Noureldn and his research team for their valuable guidance.

References

1. O. Menéndez, M. Pérez, and F. Auat Cheein, Visual-based positioning of aerial maintenance platforms on overhead transmission lines, *Applied Sciences*, vol. 9, no. 1, 165, January 2019.

2. P. V. Klaine, J. P. B. Nadas, R. D. Souza, and M. A. Imran, Distributed drone base station positioning for emergency cellular networks using reinforcement learning, *Cognitive Computation*, vol. 10, 5, 790–804, 2018.

3. F. Al-Turjman, QoS–aware data delivery framework for safety-inspired multimedia in integrated vehicular-IoT, *Elsevier Computer Communications Journal*, vol. 121, 33–43, 2018.

4. S. Choudhury, and F. Al-Turjman, Dominating set algorithms for wireless sensor networks survivability, *IEEE Access Journal*, vol. 6, no. 1, 17527–17532, 2018.

5. G. J. Lim, S. Kim, J. Cho, Y. Gong, and A. Khodaei, Multi-UAV pre-positioning and routing for power network damage assessment, *IEEE Transactions on Smart Grid*, vol. 9, no. 4, 3643–3651, July 2018.

6. F. Al-Turjman, A. Alfagih, H. Hassanein, and M. Ibnkahla, Deploying fault-tolerant grid-based wireless sensor networks for environmental applications, in *Proceedings of the IEEE International Workshop on Wireless Local Networks (WLN)*, Denver, CO, 2010, pp. 731–738.

7. M. L. Cherif, J. Leclère, and R. J. Landry, Loosely coupled GPS/INS integration with snap to road for low-cost land vehicle navigation: EKF-STR for low-cost applications, in *IEEE/ION Position, Location and Navigation Symposium (PLANS)*, Monterey, CA, April 2018.

8. F. Al-Turjman, E. Ever, and H. Zahmatkesh, Small cells in the forthcoming 5G/IoT: Traffic modelling and deployment overview, *IEEE Communications Surveys and Tutorials*, 2018. doi:10.1109/COMST.2018.2864779.

9. D. Tazartes, An historical perspective on inertial navigation systems, in *International Symposium on Inertial Sensors and Systems (ISISS)*, Laguna Beach, CA, February 2014.

10. D. Rautu, R. Dhaou, and E. Chaput, Crowd-based positioning of UAVs as access points, in *2018 15th IEEE Annual Consumer Communications & Networking Conference (CCNC)*, Las Vegas, NV, 2018, pp. 1–6.

11. F. Al-Turjman, Fog-based caching in software-defined information-centric networks, *Elsevier Computers & Electrical Engineering Journal*, vol. 69, no. 1, 54–67, 2018.

12. V. Sharma, D. N. K. Jayakody, and K. Srinivasan, On the positioning likelihood of UAVs in 5G networks, *Physical Communication*, vol. 31, 1–9, December 2018.

13. K.-W. Chiang, A. Noureldin, and N. El-Sheimy, The utilization of artificial neural networks for multi-sensor system integration in navigation and positioning instruments, *IEEE Transactions on Instrumentation and Measurement*, vol. 55, 5, 1606–1615, April 2003.

14. F. Al-Turjman, and S. Alturjman, Confidential smart-sensing framework in the IoT era, *The Springer Journal of Supercomputing*, vol. 74, no. 10, 5187–5198, 2018.

15. S. Alabady, and F. Al-Turjman, Low complexity parity check code for futuristic wireless networks applications, *IEEE Access Journal*, vol. 6, no. 1, 18398–18407, 2018.

16. H. Sun, X. Chen, Q. Shi, and M. Hong, Learning to optimize: Training deep neural networks for interference management, *IEEE Transactions on Signal Processing*, vol. 66, no. 20, 5438–5453, 2018.

17. F. Al-Turjman, H. Hassanein, S. Oteafy, and W. Alsalih, Towards augmenting federated wireless sensor networks in forestry applications, *Personal and Ubiquitous Computing*, vol. 17, no. 5, 1025–1034, June 2013.
18. F. Al-Turjman, and S. Alturjman, 5G/IoT-enabled UAVs for multimedia delivery in industry-oriented applications, *Springer's Multimedia Tools and Applications Journal*, 2018. doi:10.1007/s11042-018-6288-7.
19. M. Z. Hasan, and F. Al-Turjman, Analysis of cross-layer design of quality-of-service forward geographic wireless sensor network routing strategies in Green Internet of Things, *IEEE Access Journal*, vol. 6, no. 1, 20371–20389, 2018.
20. F. Al-Turjman, and M. AbuJubbeh, IoT-enabled smart grid via SM: An overview, Elsevier Future Generation Computer Systems, vol. 96, no. 1, 579–590, 2019.

Chapter 7

Grid-Based UAV Placement in Intelligent Transportation Systems

Fadi Al-Turjman, Sinem Alturjman, and Jehad Hamamreh

Antalya Bilim University

7.1 Introduction

Unmanned aerial vehicles (UAVs)-based projects in the smart city paradigm are no more a slim chance. It is a present moment and has expansive extension monetary and social chance. Crisis reaction and coordination, open or private infrastructure control, coordination or support to astute transportation systems are a portion of spaces in which UAV may end up key in the city of things very soon. Moreover, these flying sensors will turn into a critical component of urban monitoring systems. Thus, there exists a need for merging their traditional sensors that gather data with the existing/forthcoming wireless communication technologies [1,2].

UAVs have pulled in a critical consideration in an assortment of fields, particularly in intelligent transportation systems (ITS). UAVs speak of a vital help to the current advancements to control street traffic and screen occurrences [3]. Attributable to their three dimensional (3D) portability, UAVs have better opportunities in rush-hour monitoring, and security enhancement, contrasted with typically settled roadside units (RSUs). That is because UAVs can cover larger territories with no limitation identified by the street arrangement. In the end, by utilizing UAVs, distinctive capacities of ITS will be executed at a lower cost. In addition, UAVs can proficiently perform the asked tasks without being constantly connected via preprogramming methods that decrease vitality and costs [4]. For instance, UAVs can fly around the area of occurrence to give fundamental help, or if nothing else, to send back a report about a circumstance [5]. What's more along these lines, UAVs can act like flying mishap report specialists, flying relay nodes, flying imagery units, flying police eyes, and flying commuter traffic signs. On account of a flying mishap report operator, and as delineated in Figure 7.1, one UAV can travel to the mishap area and provide a report at that point and transmit its report through different UAVs by means of device-to-device multihop interchanges. Another adjacent UAV or an RSU that approaches the system would then be able to forward the answer to the significant element [3]. UAVs may be an extraordinary elective-utilizing picture, preparing calculations that recognize streets and vehicles status all the while [6,7]. The rescue group can be provided with a UAV that can rapidly fly over the traffic to the mishap area. This UAV can send a nitty-gritty report about the circumstance, in the wake of achieving the scene. It can likewise be utilized to build up a constant correspondence channel (voice and video) between the mishap place and the safeguard group. Moreover, current ITS projects, which significantly rely on the city cellular infrastructure, work effectively under normal circumstances. However, in events of natural disasters, such systems are relatively fragile and can easily be broken. In such circumstances, UAVs can assist as a substitute to provide structureless communications framework for communicating emergency and safety information in ITS. To make these UAVs feasible, it is essential to assure their cost, energy efficiency, and reliable operation within restrictive and harsh outdoor operational conditions. It is also needed to guarantee secure and correct transmission of information in mission critical applications while cooperating with existing sensory frameworks, forming what is called the flying ad hoc networks (FANETs).

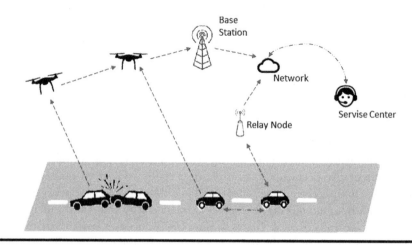

Figure 7.1 UAVs can be used as a flying relay node to deliver recorded traffic events to a service center.

In this work, we investigate the integration of UAVs and future communication networks, such as 5G as part of the existing internet of things (IoT) and cloud platforms. We aim at proposing an optimized UAV placement that assures the aforementioned characteristics. One of the major drawbacks of UAV is that its range is limited and that's due to the battery capacity, as its endurance is restricted. Another challenge is the associated high price. That is why there's a need to plan and optimize the number and locations of drones while being connected and covering a specific region. Therefore, we further explore this connectivity problem while considering practical issues. Given an area where the service has to be covered, we optimize the determination of where to locate the UAV so that each segment of interest (SoI) in the selected area is covered. We aim at minimizing the total used UAV count, which can lead to reduced cost while providing coverage, connectivity, and reliable communication links.

7.2 Related Work

Recent developments in microelectromechanical systems (MEMS) technology and very large-scale integration have been influential in transforming large base station (BS) systems (BSS) to minute structures, which enables the adaptation of small-sized drones (UAVs). The UAVs are capable of replicating technology features of a BSS and can be used to form a small coverage area. Drones, with the ability to move autonomously and to hover over the affected area, can function as a small cell to establish communication with the user equipment (UE) active in the designated emergency coverage area. Hypothetically, with the presence of sufficient drones, the communications outage area in vulnerable regions can be fully covered. The restoration of a communication network in such areas using drones provides a rapid and reliable alternative to reconfigure and replicate necessary functionalities of the affected BSS. These drone small cells (DSCs) can also be used to enhance and extend communications coverage in disaster areas where on-ground repairs are not feasible. The ability of DSCs to reposition itself and respond to UE by reducing distance extends coverage, decreases outage probability of the UE in coverage zones, improves bandwidth efficiency, and optimizes system throughput. Research in DSC is still in its infancy, and many practitioners and academics are keen to pursue their research in this scholarly area.

Several research works were conducted to address further challenges in deployed UAVs for critical applications. In Ref. [8], UAVs are proposed as part of a system targeting postdisaster scenarios. The running systems on each UAV are clarified and assessed using a prototype helicopter. The long-term evolution-unlicensed (LTE-U) technology is proposed in Ref. [9] for DSCs to enhance the achievable broadband throughput to aid postdisaster scenarios. An ON/OFF game-based mechanism is employed for effective use of LTE-U and to reach a correlated equilibrium. Numerical simulations are employed in Ref. [10] to study the coverage that can be provided by UAV-based BSs. The study attempts to minimize the number of stops and amount of delays for a single UAV that needs to visit various positions to completely cover the potential disaster area. This study is further extended in Ref. [11] for multiple UAVs. A framework is proposed for optimizing 3D placement and the mobility of UAVs. Simulations performed using MATLAB® provide results that show significant enhancements using the proposed approach, especially in terms of reductions in transmission power of IoT devices and system reliability. Through these results, the significance of intelligent decisions in moving and deploying UAVs has been emphasized. A drone cooperation scenario is considered in Ref. [12]. The UAV-based BSs are employed together with conventional BSs in an attempt to aid disaster-struck regions where terrestrial infrastructure is damaged. The main focus of this study is efficient power allocation strategies for the microwave BS as well as smaller UAV-based BSs. The power control strategy presented is self-adaptive depending on the interference threshold employed as well as data rate requirements. Factors such as UAV altitude and number of ground users are considered with an analytical abstraction for simulations. The importance of incorporation of drones in the multitier heterogeneous networks for better network coverage and capacity is emphasized in this study as well. Authors in Ref. [13] talk about finding an optimal position for UAV, such that the sum of time durations of the uplink transmissions is maximized.

The decision-making and evaluation processes of these studies are mainly dependent on high-level analytical abstractions of specific scenarios considered. We believe that there are factors above the physical and data link layers that can affect the optimization of heterogeneous infrastructures. Moreover, considering UAV deployment based on uplink and downlink resources and optimization using various methods based on game theory may not be sufficient in mission critical ITS applications, where significant design factors such as outdoor conditions, mobility-related issues, and availability of other facilities should also be considered together with facilities provided by UAVs. Furthermore, considering the limited fly times mainly due to limited energy resources of UAVs, the optimum configuration for the transmission of safety critical information becomes even more critical. In this line, authors in Ref. [14] introduce a minimum cost drone location problem. In their work, they use a two-dimensional terrain to find the optimal location and number of drones to observe given targets, which could be mobile or static in a given region. Authors develop an integer linear and a mixed integer nonlinear optimization equation by considering the coverage of drones and the energy consumed. Authors in Ref. [15] look into recent brainstorm optimization algorithm, to find the optimal position of locating static drones in a monitored area such that coverage is maximized. The authors tested the proposed method with uniformly and clustered deployed targets. Upon reviewing their results, we can conclude that their brainstorm optimization is not appropriate for solving the targeted UAV placement problem under outdoor operational conditions. In Ref. [16], the authors compare between optimization methods and heuristic algorithms, such as simulated annealing (SA) and genetic algorithms (GAs). The authors conclude that the heuristic algorithms are potentially better for dynamic scenarios. However, optimization methods are ideal for static UAVs and highly demanded applications that would function under stress. The authors in Ref. [17] propose a new heuristic method called modified genetic and SA algorithm. They present simulation results and conclude that the proposed method generates high-quality results compared with pure GA. In Ref. [18], the authors compare the results obtained from pure GA and SA algorithm, where the problem of optimizing the topological design of FANET is tackled. The authors conclude that the average cost of GA solutions is less than the average cost of SA solutions.

Spurred by the advantages of optimization and graph theory, and additionally, the 3D virtual grid, this chapter offers a UAV-optimized deployment method for provisioning FANETs with extremely connected backbone while considering network cost and lifetime limitations in ITS applications. For more productivity and dissimilar to other deployment methods, we are finding the most conceivable grid vertices to be scanned for ideal UAV placement as opposed to searching an interminable 3D space.

The organization of this chapter goes as follows. The considered FANET, communication, and lifetime models along with the problem definition are presented in Section 7.3. The proposed UAV deployment strategy is explained in Section 7.4. Then, performance evaluation results using simulations are exhibited and interpreted in Section 7.5. Finally, a concise conclusion is proposed in Section 7.6.

7.3 Assumed Models and Problem Definition

Considered FANET models are discussed in this section. A graph-based topology is assumed to model the network and mathematically represent its connectivity. In addition, we detail the considered lifetime, cost, and wireless communication models.

7.3.1 FANET Model and Problem Definition

In this study, a flat FANET is considered. It comprises of UAVs that have transmission ranges equal to r and talks occasionally to the BS to convey the caught pictures in an ITS application. The FANET topology is represented by a graph $G = (V, L)$, where $V = \{v_0, v_1,..., v_{nc}\}$ is the arrangement of n_c competitor vertices, L is the arrangement of links in graph G, and $(i, j) \in L$, if UAVs at v_i and v_j have enough probability to communicate and build up the corresponding link. We comment that the organization of UAVs in this study is autonomous of the MAC (medium access control), with a transmission rate T for every UAV

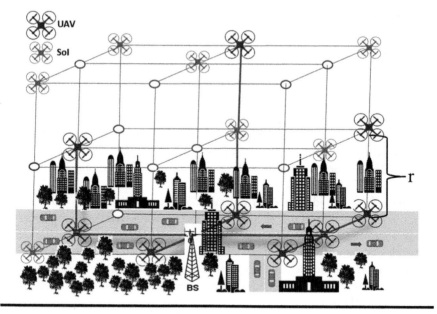

Figure 7.2   Grid model for FANET deployment in ITS.

per time unit. We also consider the S-MAC (simple medium access control) protocol to deal with MAC in this chapter with a half obligation cycle and just 6 bytes control field length in the traded packets for more vitality. In addition, we expect that the influx rate of pictures is following Poisson arrival rate, which is typical in FANET reenactments.

Figure 7.2 delineates the assumed virtual grid in this chapter, where the lattice edge length should be equivalent to a UAV transmission extend r. It is expected that all UAVs have equal transmission extend r. In this grid, UAVs are set initially close to the observed SoI. Additional UAVs are later positioned on the proper vertices connected to the biggest number of SoI in the city. The BS is positioned according to application prerequisites, and it is the information sink for FANET. Consequently, we can define the considered general UAV deployment issue in ITS as follows:

Considered Problem Definition: Given a specific set of SoI and BS with predefined positions, find locations of UAVs that maximize the connectivity between themselves and the BS while cost and lifetime conditions are satisfied.

7.3.2 Communication and Cost

Since UAVs are the most expensive component in the assumed FANET, cost is modeled in this chapter by the count of UAVs hovering in the targeted site.

For simplicity, we consider a similar cost for used UAVs. As for the wireless communication, we assume a probabilistic one, in which remote signals rot with separation, as well as are lessened and influenced by obstacles. In a like manner, the communication scope of each UAV must be represented by a discretionary shape according to surrounding environmental conditions. This model can be depicted as follows [19]:

$$P_r = K_0 - 10\gamma \log(d) - \mu d \qquad (7.1)$$

where P_r is the gotten signal strength, d is the Euclidian separation between sender and recipient, γ is the path loss exponent, μ is an irregular variable that pursues a log-normal distribution with zero mean and difference δ^2 to depict signal lessening impacts in the observed segment, and K_0 is a constant dependent on the sender, recipient, and observed region heights.

For simplicity and clarity in the presentation of our proposed scheme, we assume a simplified channel model, similar to the one used in Ref. [19], as our contribution in this work is not related to channel modeling but rather on the optimization scheme. However, it should be pointed out that one can consider using a much more accurate channel model for UAV communication, such as the ones proposed in Refs. [20] and [21]. In particular, the work in Ref. [20] experimentally obtained a channel model for path loss exponents along with shadowing for the radio channel between UAV and BS cellular tower. The empirical results indicate that path loss reduces as the UAV goes higher, up to a height around 100 m, at which UAV experiences a perfect free-space propagation model. More precisely, the parameters describing the propagation channel of UAVs are found to be height-dependent.

7.3.3 Lifetime Model

Lifetime models vary in the manner in which they view a FANET as it is still operational. These models can depend on the availability of UAV or on the level of functioning UAVs, which have enough vitality to achieve their appointed assignments. Nonetheless, losing a couple of UAVs may not fundamentally influence the general FANET execution, particularly when excess UAVs are utilized. For the most part, in ITS applications, a few UAVs are relegated to screen a specific criterion in the system. Therefore, the idea of UAV repetition ought to be tended to. What's more, lifetime models depending on the previously mentioned definitions don't think about the UAV assignment that could be catching pictures, relaying information packets, or speaking to a gateway. Hence, we introduce an ITS-particular lifetime designation.

ITS-specific Lifetime Designation: FANET lifetime is defined by the amount of time period, since the positioning of UAVs till the time a percentage of UAVs are connected reach a minimum predefined threshold τ.

7.4 UAV Deployment Strategy

The UAV positioning issue proposed in this chapter boundlessly has a huge search space, and finding the ideal arrangement is exceptionally a nonpolynomial problem. Hence, we propose a 3D lattice-based deployment that confines the pursuit space to a progressively reasonable size. We consider the 3D landscape of the observed site before starting deployment optimization. Accordingly, potential coordinates on the grid are predecided; in a way, nondoable coordinates are prohibited from the inquiry space. We use these coordinates to put on our placement strategy in two stages. The initial stage called first phase is utilized to put the base count of UAVs on the grid to set up a connected FANET. The second stage, called second phase, is utilized to pick the ideal positions for additional UAVs required to amplify the FANET availability/connectivity with limitations on expenses budget and required lifetime period. This two-stage positioning methodology is named as optimized 3D placement with lifetime and cost conditions (O3DwLC).

7.4.1 First Phase of O3DwLC

The main stage in O3DwLC is accomplished by developing an associated backbone (B) utilizing first-phase UAVs called FPUAVs. Positions of these UAVs are enhanced so as to utilize the base B that can cover and associate SoI to the BS. Consequently, we put on the minimum spanning tree (MST) as depicted in Algorithm 1.

Algorithm 1: MST to Construct the Connected Backbone B

1. **Function ConstructB** (IS: Initial Set (IS) of nodes to construct B)
2. **Input:**
3. A set *IS* of the SoIs and BS positions.
4. **Output:**
5. A set of connected UAVs (CU) in the FANET, with least UAVs, and a BS coordinate forming B.
6. **begin**
7. CU = set of closest two UAVs in IS;
8. $CU = CU \cup$ least UAVs to connect them;
9. $IS = IS - CU$;
10. N_d = number of residual *IS* UAVs not in *CU*;
11. $i = 0$;
12. **for each** residual UAV_i in *IS* **do**
13. Compute M_i: Positions of least count of UAVs needed to connect UAV_i with the nearest UAV in CU[1].

---

[1] This is realized by calculating the least count of nearby grid vertices, which build up a path from the disconnected SoI at v_i to *CU*.

14. $i = i + 1$;
15. **end**
16. $M = \{ M_i \}$
17. **while** $N_d > 0$ **do**
18. $SM =$ Smallest M_i;
19. $CU = CU \cup SM \cup UAV_i$;
20. $IS = IS - UAV_i$;
21. $M = M - M_i$;
22. $N_d = N_d - 1$;
23. **end**
24. **end**

Algorithm 1 goes for building the MST utilizing the cubic grid competitor positions. Line 7 of Algorithm 1 looks for the nearest two UAVs in the underlying set IS, which has the SoI and BS. On the off chance that the two UAVs are not contiguous on the grid (i.e. $Pc \leq \tau$), it counts at line 8 the least number of vertices on which the UAVs must be set to set up a way between these two UAVs. Next to connecting UAVs in a set called CU, we iteratively search for the following nearest vertex that must be associated with the CU. This is accomplished through lines 12–22 in Algorithm 1. For more explanation on positioning the FPUAVs, we build up the following example.

Example 1

Let's say we have seven SoIs prepositioned with one BS on the assumed grid setup, as illustrated in Figure 7.3a. At that point, we need to look for the least count of UAVs, represented by N_{MST} to associate these SoI with the BS, as depicted in Figure 7.3b. Places of N_{MST} UAVs are dictated using Algorithm 1. This algorithm first builds the IS comprising of a UAV for each SoI and BS setup. At that point, a CU set is started by the nearest two UAVs in IS, which are UAVs at vertices 15 and 17, in this case. These directions are then expelled from IS. Clearly, by including just a single UAV at vertex 14, UAVs at 15 and 17 wind up associated. Thus, the position of that UAV is included in the set CU and residual count of vertices N_d in is set to be 6. Next, we figure M_i for the remaining UAVs at vertices 1, 5, 19, 23, 25, and 27 in IS, which can have 2, 0, 2, 0, 1, and 1 as potential UAV positions. Since the set M_1, related with the BS set at vertex 5, has the least count of required positions, we place M_1 in the set SM and CU ends up equivalent to $\{15, 17, 14, 5\}$. M and IS are then refreshed and N_d is decremented by 1. By rehashing this procedure until the point that N_d is equivalent to 0, we acquire the last associated CU that appeared in Figure 7.3b, where UAVs are the FPUAVs of a remote FANET to convey. The sent FPUAVs with BS build up the FANET backbone.

Availability of B produced in this stage of the plan is estimated by assuming B as a graph with a Laplacian matrix $L(B)$, as depicted in Figure 7.4 [22]. Given $L(B)$, B connectivity is mathematically modeled by calculating the second smallest

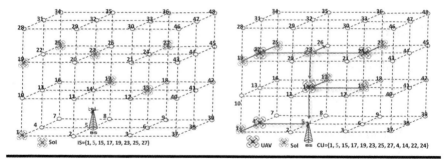

Figure 7.3 An example of FPUAV placement using Algorithm 1. (a) Before applying Algorithm 1 (i.e. without FPUAVs) and (b) after applying Algorithm 1 (i.e. with FPUAVs). *r* is the grid edge length.

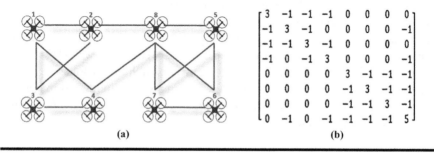

Figure 7.4 (a) A graph with 8 vertices (or UAVs) and 13 edges. The graph's connectivity characteristics are as follows: one vertex to disconnect, two edges to disconnect. Laplacian matrix of this graph is shown in (b) and $\lambda_2 = 0.6277$.

eigenvalue λ_2, where λ_2 represents the least count of UAVs and links, whose removal would partition B. By seeking the largest value of λ_2, we increase the needed count of UAVs and edges between them to partition data routes in the FANET backbone. Hence, more reliable[2] UAV network can be realized due to the ability to overcome significant topology changes caused by undesired wireless communication conditions and UAV failures [14]. Accordingly, extra UAVs called second-phase UAVs (SPUAVs) are positioned in the second stage of O3DwLC approach.

7.4.2 Second Phase of O3DwLC

In this stage, we prioritize the required positions of SPUAVs such that λ_2 of $L(B)$ found in the first stage is maximized. We begin by maximizing λ_2. Considering n_c potential locations for SPUAVs, we choose the best N_{SPUAV} UAV amongst these n_c

---

[2] Reliability here is defined by the existence of a functional path from all SoI to BS while experiencing UAVs and link failure.

Table 7.1   Notations Used in the Placement Problem

Notation	Description
α_i	A Boolean value equals 1 when UAV$_i$ is positioned on vertex i and 0 otherwise
A_i	Matrix that results by positioning UAV$_i$ on vertex i
n	Summation of $N_{UAV} + N_{MST} + 1$
L_i	Laplacian matrix produced at the beginning by the positioned FPUAVs and BS
$I_{n \times n}$	Identity matrix of dimension $n \times n$

in terms of connectivity. Consequently, and based on Table 7.1, we can formulate the optimization problem as follows:

$$\max \lambda_2 \left(L(\alpha) \right) \tag{7.2}$$

$$s.t. \sum_{i=1}^{n_c} \alpha_i = N_{SPUAV}, \ \alpha_i \in \{0, 1\},$$

where

$$L(\alpha) = L_i + \sum_{i=1}^{n_c} \alpha_i A_i A_i^T \tag{7.3}$$

In any way, solving Eq. (7.2) will take a very long time, especially for substantial n_c values in extensive-scale ITS applications. This is because of the included calculations required for discovering λ_2 for a large number $\left(= \begin{pmatrix} nc \\ N_{SPUAVs} \end{pmatrix} \right)$ of Laplacians. In this way, we require a computationally effective technique to settle Eq. (7.2), notwithstanding constrained pursuit space that diminishes n_c. Also, subsequently, we regenerate it as a standard semidefinite program (SDP) [[2,22]].

By replacing the binary constraint $\alpha \in \{0, 1\}$ by $\alpha \in [0, 1]$, we can reformulate Eq. (7.2) as convex with linear constraint problem [2]:

$$\max \lambda_2 \left(L(\alpha) \right) \tag{7.4}$$

$$s.t. \sum_{i=1}^{n_c} \alpha_i = N_{SPUAV}, \ 0 \le \alpha_i \le 1,$$

So as to embed the FANET's lifetime requirement into Eq. (7.4), let B be functional at the beginning for a number of rounds equal to initial rounds (IRs). Consider including one UAV of the SPUAVs set into B would extend the lifetime by additional rounds equal to ER_i. At that point, to ensure that the FANET will remain operational for the least count of rounds RLT, the additional and initial B rounds must be more noteworthy than or equivalent to RLT, as explained in the following inequality:

$$-\sum_{i=1}^{n_c} ER_i \leq (IRs - RLT) \qquad (7.5)$$

Since the cutoff condition of FANET lifetime definition is used in ascertaining both ER_i and IRs values, Eq. (7.5) reflects the ITS-particular lifetime limitation in our proposed O3DwLC approach. From Eqs. (7.4) and (7.5), SPUAV positions that amplify λ_2 with imperatives on lifetime and cost are determined by

$$\max S \qquad (7.6)$$

$$s.t. \ S\left(\boldsymbol{I}_{nxn} - \frac{1}{n}11^T\right) \leq L(\alpha), \ \sum_{i=1}^{n_c} \alpha_i \leq N_{SPUAV},$$

$$-\sum_{i=1}^{n_c} ER_i \alpha_i \leq (IRs - RLT), \ 0 \leq \alpha_i \leq 1,$$

Algorithm 2 given later condenses the second-stage UAV's organization proposed in this study, where the inquiry space is constrained to n_c matrix vertices near FPUAVs, which we refer to as set P. And thus, we are further limiting our problem search space without influencing its correctness. For a better understanding of Algorithm 2, see Example 2.

Algorithm 2: SPUAVs Positioning Function

> 1. **Function SPUAVs** (B: Backbone constructed by SoIs, FPUAVs, and BS, P)
> 2. **Input:**
> 3. A set B of the SoIs, FPUAVs, and BS positions.
> 4. An ideal set P of n_c candidate positions for SPUAVs.
> 5. **Output:**
> 6. A set SP of SPUAVs positions increasing connectivity of B
> 7. **begin**
> 8. **for** $(i = 1; i < n_c; i++)$
> 9. L_i = Laplacian matrix of B
> 10. **IRs** = number of rounds B can stay functional for

11. $\mathbf{A}_i$ = matrix associated to vertex i on the grid
12. $\mathbf{ER}_i$ = extra rounds reached by positioning UAV at vertex i
13. **end**
14. $\mathbf{SP}$ = SDP Solution in (6)
15. **End**

Example 2

Suppose we have up to two additional SPUAVs to amplify the availability of B produced in Figure 7.3b and guarantee no less than 20 rounds lifetime. For this situation, $N_{SPUAV} = 2$, $n = 12$, and $RLT = 20$. We begin by deciding the set P to indicate our pursuit space in this stage. Then, we decide the underlying Laplacian L_i with respect to B using Eq. (7.3). As part of that, we also calculate the IRs of B, which can be assumed in this example equal to 10. Based on Table 7.1, we fix α_i to 1 and calculate A_i and ER_i for the ith entry in P, where $P = \{10, 13, 16, 2, 8, 20, 26, 6, 18\}$, as depicted in Figure 7.3b. Next, we solve the SDP in Eq. (7.6). Accordingly, we will find that the most elevated two estimations of λ_2 are coming from the vertices 10 and 26. By placing two SPUAVs at these two vertices, we guarantee that the FANET lifetime will not be less than 20 rounds.

7.5 Performance Evaluation

Performance of the proposed O3DwLC under brutal operational conditions is investigated in this section. We contrast our technique with a focused approach called the shortest path 3D placement (SP3D). The SP3D method is normally used in outdoor ITS applications due to its simplicity. Moreover, SP3D is utilized as a benchmark in this exploration because of its proficiency in keeping up a predefined lifetime and picking the base count of UAVs required in developing the FANET. In SP3D, Algorithm 1 is utilized to build the B structure of FANET. This B interfaces the predecided arrangement of SoI with the BS. At that point, further UAVs are thickly dispersed close to B to upgrade FANET connectivity. O3DwLC and SP3D methods are assessed and looked at utilizing three changed measurements:

1. FANET connectivity (λ_2): this criterion reflects FANET reliability under outdoor operational conditions.
2. Count of UAVs: this indicates the FANET cost in ITS applications.
3. Total rounds: this indicates the sum of rounds the FANET can stay functional for. It reflects the FANET's lifetime.

Two primary parameters are used in this study to assess the aforeselected metrics: (1) probability of node failure (PNF) and (2) probability of disconnected nodes (PDNs). These parameters have been chosen due to their ability in reflecting the brutality dimension of the observed ITS site as far as weak signals and physical UAV damages are expected.

7.5.1 Simulation Setup

Simulations of O3DwLC and SP3D methods are performed on 500 arbitrarily created FANET topologies so as to get measurably steady outcomes in this chapter. For every topology, we apply an arbitrary UAV/edge failure and evaluate the previously mentioned metrics. Considered space dimensions are 800 by 800 by 300 (m³). Twenty irredundant SoIs notwithstanding 1 BS are arbitrarily situated on a 3D virtual grid. We accept a FANET in which each UAV is required to be operational for 20 rounds as a base lifetime condition with utmost 60 UAVs as the greatest cost confinement.

Assuming a city with dense obstacles, we set our simulation parameters as depicted in Table 7.2. Values of N_{MST} and *IRs* vary depending on the aftereffects of the first stage arrangement. We consider settled/fixed transmission power, and an edge length is equivalent to 100 m. For verification purposes, we change PNF and PDN from 0% to 60%, while holding confidence interval close to 5% at a 95% certainty level.

7.5.2 Simulation Results

While O3DwLC optimizes the coordinates of SPUAVs to provide the maximum λ_2, SP3D aims at locating additional UAVs by searching vertices that can increase the FANET structure strength as aforementioned.

As depicted in Figure 7.5, O3DwLC outperforms SP3D in terms of cost and connectivity. This figure shows λ_2 values for the compared two methods while utilizing varying counts of UAVs, and settled SoI counts to a value equal to 20 and PDN equals to 0.2. Obviously, an increment in the deployed UAVs count can lead to enhancement in connectivity while experiencing 20% isolated UAVs with

Table 7.2 Parameters of the Simulated FANET

Parameter	Value	Parameter	Value
RLT	20 (round)	N_{SPUAV}	0–60 (UAVs)
t_{round}	24 (h)	PNF	0%–60%
PDN	0%–60%	N_{SoI}	20
τ	70%	L	512 (bits)
n_c	110 (vertex)	PNF	10%–60%
γ	4.8	T	100 (packet/round)
δ_2	10	P_r	−104 (dB)
K_0	42.152	r	100 (m)

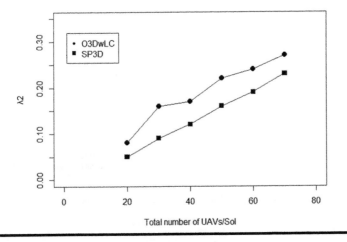

Figure 7.5   Connectivity vs. the deployed count of UAVs.

O3DwLC approach. Furthermore, Figure 7.5 depicts the efficiency of generated FANET topologies by O3DwLC in terms of cost while satisfying a specific connectivity level. For example, by deploying just 30 UAVs, O3DwLC can provide higher λ_2 than what SP3D can provide with 70 UAVs.

Moreover, the influence of PDN on the generated FANET lifetime while assuming both approaches has been checked in Figure 7.6. The count of UAVs in this comparison is set to 40. From Figure 7.6, we conclude that the FANET produced by O3DwLC can continue functioning for extended periods than the FANET generated by SP3D. This implies the desired reliability feature even while experiencing PDN equal to 60%. Figures 7.5 and 7.7 depict that the produced FANET lifetime can be increased by increasing its λ_2. In Figure 7.7, the lifetime

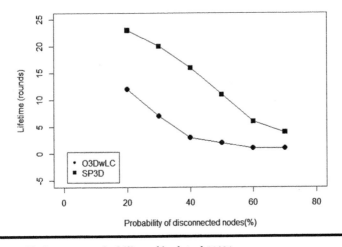

Figure 7.6   Lifetime vs. probability of isolated UAVs.

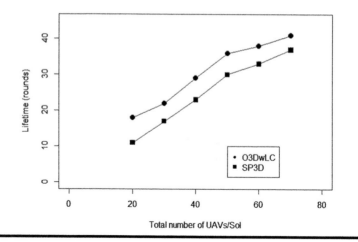

Figure 7.7   Lifetime vs. number of UAVs under *PDN* = 0.2 and *RLT* = 20.

influence of FANET on O3DwLC and SP3D performance has been investigated, while setting the PDN parameter to be equal to 20%. Apparently, O3DwLC outperforms the SP3D method in terms of rounds count, which is a critical issue in outdoor ITS applications. Also we remark that different generated FANETs by the two deployment approaches experience decreasing lifetime difference as the UAV count increases. This can be returned to the mass growth of UAVs, which makes the generated FANETs more connected and tougher to divide.

Figures 7.8 and 7.9 depict that O3DwLC approach beats the SP3D while experiencing significant PDN and PNF values. FANETs produced by O3DwLC continue to be tightly attached even though a value equal to 50% and PNF equal to 50% value is assumed. Indeed, this behavior is strongly recommended in ITS

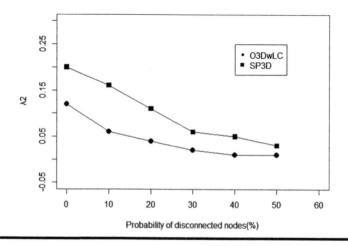

Figure 7.8   Connectivity vs. probability of disconnected UAVs.

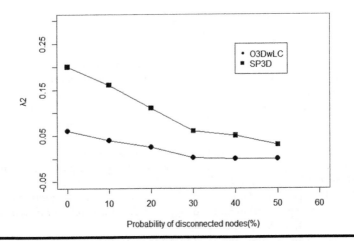

Figure 7.9   FANET connectivity vs. probability of CHs/RNs failure.

applications which are typically functioning under harsh outdoor conditions. Nevertheless, when the probability of UAV/link failure reaches a particular value, FANET's availability degrades severely. Consequently, such probability of failure is of utmost importance in UAV deployment planning, especially during premature phases to improve the FANET behavior in practice.

7.6 Conclusion

The problem of UAV placement in FANETs utilized by ITS applications is investigated. We aimed at maximizing FANETs availability while guaranteeing specific lifetime periods and cost-effectiveness. To address the targeted problem complexity, we propose a two-phase UAV placement approach in 3D space using MST and SDP techniques. Towards further practicality, ITS-specific lifetime and communication models are assumed. It leads to practicality in produced UAV plans. Extensive simulations and experimental results have been provided under harsh operational circumstances. It indicates that our proposed O3DwLC method can offer strongly connected FANETs and practically guaranteed lifetime for ITS. Furthermore, the presented O3DwLC approach offers a reasonable benchmark for FANETs in ITS applications.

References

1. F. Al-Turjman, and S. Alturjman, 5G/IoT-enabled UAVs for multimedia delivery in industry-oriented applications, *Springer's Multimedia Tools and Applications Journal*, 2018. doi:10.1007/s11042-018-6288-7.

2. F. Al-Turjman, and S. Alturjman, Context-sensitive access in industrial internet of things (IIoT) healthcare applications, *IEEE Transactions on Industrial Informatics*, vol. 14, no. 6, 2736–2744, 2018.
3. L.-Y. Yu, and L. Lu, Research on test data generation based on modified genetic and simulated annealing algorithm, in *8th International Conference on Supply Chain Management and Information Systems (SCMIS), 2010*, IEEE, 2010, pp. 1–3.
4. F. Al-Turjman, QoS–aware data delivery framework for safety-inspired multimedia in integrated vehicular-IoT, *Elsevier Computer Communications Journal*, vol. 121, 33–43, 2018.
5. M. Z. Hasan, and F. Al-Turjman, Analysis of cross-layer design of quality-of-service forward geographic wireless sensor network routing strategies in green internet of things, *IEEE Access Journal*, vol. 6, no. 1, 20371–20389, 2018.
6. D. R. Thompson, and G. L. Bilbro, Comparison of a genetic algorithm with a simulated annealing algorithm for the design of an ATM network, *IEEE Communications Letters*, vol. 4, no. 8, 267–269, 2000.
7. F. Al-Turjman, E. Ever, and H. Zahmatkesh, Small cells in the forthcoming 5G/IoT: Traffic modelling and deployment overview, *IEEE Communications Surveys and Tutorials*, 2018. doi:10.1109/COMST.2018.2864779.
8. F. Al-Turjman, Cognitive routing protocol for disaster-inspired internet of things, *Elsevier Future Generation Computer Systems*, vol. 92, 1103–1115, 2019.
9. F. Al-Turjman, C. Altrjman, S. Din, and A. Paul, Energy monitoring in IoT-based Ad Hoc networks: An overview, Elsevier Computers & Electrical Engineering Journal, vol. 76, 133–142, 2019.
10. F. Al-Turjman, H. Hassanein, and M. Ibnkahla, Towards prolonged lifetime for deployed WSNs in outdoor environment monitoring, *Elsevier Ad Hoc Networks Journal*, vol. 24, no. A, 172–185, January 2015.
11. G. Tuna, B. Nefzi, and G. Conte, Unmanned aerial vehicle-aided communications system for disaster recovery, *Journal of Network and Computer Applications*, vol. 41, 27–36, 2014.
12. D. Athukoralage, I. Guvenc, W. Saad, and M. Bennis, Regret based learning for UAV assisted LTE-U/WiFi public safety networks, in *2016 IEEE Global Communications Conference (GLOBECOM)*, IEEE, 2016, pp. 1–7.
13. D. Zorbas, L. D. P. Pugliese, T. Razafindralambo, and F. Guerriero, Optimal drone placement and cost-efficient target coverage. *Journal of Network and Computer Applications*, vol. 75, 16–31, 2016.
14. A. Ghosh and S. Boyd, Growing well-connected graphs, in *Proceedings of the IEEE Conference on Decision and Control*, San Diego, CA, 2006, pp. 6605–6611.
15. S. Alabady, and F. Al-Turjman, Low complexity parity check code for futuristic wireless networks applications, *IEEE Access Journal*, vol. 6, no. 1, 18398–18407, 2018.
16. E. Tuba, R. Capor-Hrosik, A. Alihodzic, and M. Tuba, Drone placement for optimal coverage by brain storm optimization algorithm, in *International Conference on Health Information Science*, Springer, Cham, 2017, pp. 167–176.
17. H. Shakhatreh, and A. Khreishah, Optimal placement of a UAV to maximize the lifetime of wireless devices, 2018. arXiv preprint arXiv:1804.02144.
18. A. Rodriguez, A. Gutierrez, L. Rivera, and L. Ramirez, RWA: Comparison of genetic algorithms and simulated annealing in dynamic traffic. In *Advanced Computer and Communication Engineering Technology*, pp. 3–14. Springer, Cham, 2015.

19. M. Mozaffari, W. Saad, M. Bennis, and M. Debbah, Unmanned aerial vehicle with underlaid device-to-device communications: Performance and tradeoffs, *IEEE Transactions on Wireless Communications*, vol. 15, no. 6, 3949–3963, 2016.
20. M. Mozaffari, W. Saad, M. Bennis, and M. Debbah, Mobile unmanned aerial vehicles (UAVs) for energy-efficient internet of things communications. *IEEE Transactions on Wireless Communications*, vol. 16, no. 11, 7574–7589, 2017.
21. S. A. R. Naqvi, S. A. Hassan, H. Pervaiz, and Q. Ni, Drone-aided communication as a key enabler for 5G and resilient public safety networks. *IEEE Communications Magazine*, vol. 56, no. 1, 36–42, 2018.
22. S. Boyd, Convex optimization of graph Laplacian eigenvalues, *Proceedings of the International Congress of Mathematicians*, vol. 3, no. 63, 1311–1319, November 2006.

Chapter 8

A Cost-Effective Framework for the Optimal Placement of Drones in Smart Cities

Fadi Al-Turjman, Reda Daboul, and Semail Ulgen
Antalya Bilim University

Hadi Zahmatkesh
Middle East Technical University

8.1 Introduction

In recent decade, due to the advancement in technologies, cities are being smarter, focusing to enhance the lives of its inhabitants and sustain its resources. Smart cities [1] are led by strategic administrations that support technology and innovation. The aim of smart cities is to maximize the efficient use of valuable resources to foster sustainable growth. Unmanned aerial vehicle (UAV) is considered a crucial part of smart cities. The main objective of a smart city is to improve its resident's life by providing low-cost services and efficient infrastructure. UAVs are already being used to document accident scenes [2], support first responder activities, and monitor construction sites [3], but they are ready to become an integral part of a smart city's network as well.

UAVs can be used to gather key intelligence data on movements of potential threats and to help in determining locations of threats and providing detailed topographic information in real time. They can also be utilized in providing an accurate representation of an area using images, which can help to rescue human and animals in case of a disaster [4].

The concept of smart city is converting cities into digital societies, transforming the life of its citizens to an easy life in every facet, and intelligent transport system (ITS) becomes an indispensable component among all. ITS [5] is the application of sensing, analysis, control, and communication technologies to ground transportation to improve safety, mobility, and efficiency. It includes a wide range of applications that process and share information to ease congestion, enhance traffic management, minimize environmental impact, and increase the benefits of transportation to commercial users and the public in general.

Drones are widely utilized in various areas. Their applications can be classified into environmental- and industrial-based applications. In the following subsections, we list and describe some typical ones of these applications.

8.1.1 Environmental-Based Applications

Drones or UAVs are becoming increasingly popular for monitoring the environment. The technology has entered various fieds such as surveillance and search and rescue operations. Drones can be equipped with sensors and cameras, making them ideal for monitoring environment. In this section, we briefly discuss some important applications of drones in monitoring environments.

Disaster Management: Disasters are affecting different regions of the world every year. They are unstoppable events that are either natural or man-made, such as wildfire, earthquake, terrorist attacks, and floods. One of the major challenges faced by the rescue team during an enormous disaster is to find survivors and victims as early as possible and to take them out of the disaster area to ensure that they are not stuck under the destroyed area. In this regard, drones can help to detect people in disaster areas [6]. They can be equipped with sensors and camera to identify the precise location of survivors as early as possible. The data can be sent to the rescue team for further investigation and action.

Vegetation Management: Drones provide an important innovation in vegetation monitoring. By using the right sensors and an appropriate camera attached to it, it is possible to map the health of the crops by determining soil quality, humidity, and pollution in the area. The advantage of using drone-based system for vegetation monitoring is that unlike satellite images, drones can provide more information in relatively smaller areas. Moreover, the cost of using drones is much lower compared with manned flights, and therefore, it makes the technology more accessible.

Water Resource Management: Water management is one of the main issues in agriculture, in which new technologies such as drones can provide solutions. The use of drones in water management can help to provide solution on how to manage irrigation water and maximize its efficiency. For example, integration of

UAV photogrammetry and image recognition technology can be used to solve the limitations of the existing measuring tools and techniques for water level measurements in the field [7].

8.1.2 *Industrial-Based Applications*

Drones or UAVs are playing a significant part in the industrial internet of things (IIoT) [8]. They can be valuable in industrial applications such as mining [9], oil and gas [10], and construction [11]. We briefly discuss some important aspects of these industrial applications in this section.

Mining Activities: Drones or UAVs can enhance security in applications related to mining activities with real-time information, such as latest surface surveys for enhanced blast patterns, quick and accurate pre- and postblast information, and recognizing of misfire and wall damage. Moreover, drones can provide an effective approach to monitor stores and assist with area exploration as well as general management. In addition, miners can gain benefits from the use of drones in the design of roads and dumps, as they help them to find out more efficient approaches in terms of environmental impacts.

Oil and Gas: UAVs or drones have been deployed and used by several operators in oil and gas sector for various activities in difficult environments [10]. These activities include data collection, inspection, and exploration. Using inspection drones in oil and gas sector has several advantages over traditional inspection methods. For example, it eliminates major dangers to personnel involved in traditional inspection activities in dangerous environments. Moreover, a significant reduction in cost is achieved due to ease of access in difficult environments.

Construction: Aerial craft can be used in almost every stage of the engineering process, from planning to final construction. Helicopters and airplanes are already being used in civil engineering for different purposes such as mapping from a plane and producing marketing films for tourist destinations. Utilizing drones can significantly reduce the expense and time traditionally involved in various stages of the engineering process, such as construction of roadways and forest road, and coastal erosion.

The aim of this study is to monitor the ITS with drones by monitoring a specific target or area. We propose a cost-effective framework to minimize the total number of drones required to monitor the environment while providing the maximum visionary coverage for the target.

The remainder of this article is organized as follows. Section 7.2 presents the existing studies related to the use of UAVs in the IoT era. Problem description and the proposed framework are outlined in section 7.3. Section 7.4 discusses the performance metrics, results, and findings of the study. Finally, Section 7.5 concludes this chapter. A list of abbreviations together with their brief definitions used throughout the chapter is provided in Table 8.1 to help the readers in understanding the abbreviated terms.

Table 8.1   List of Abbreviations

Abbreviated	Name
AED	Automated External Defibrillator
DSP	Drone Scheduling Problem
ECA	Emission Control Area
IIoT	Industrial Internet of Things
IoT	Internet of Things
ITS	Intelligent Transport System
LRBA	Lagrangian Relaxation-Based Approach
ODP	Optimal Drone Placement
OPA	Optimized Placement Approach
OS	Operating System
TSP	Traveling Salesman Problem
UAV	Unmanned Aerial Vehicle

8.2 Related Works

This section provides a review on existing studies related to use of drones and their deployment approaches in the IoT era. Recently, the use of UAVs or drones to control the emissions of sailing vessels has attracted too much attention due to its significant potential for performing regulations in Emission Control Areas (ECAs). In [12], the authors propose a drone scheduling problem (DSP) such that a group of planned tours is developed for drones to examine the vessels in ECAs. The dynamics of sailing vessels are modeled using a real-time location function in a deterministic manner. They also propose a Lagrangian relaxation-based approach (LRBA), which is able to gain the best solution for the problem in large-scale cases. The results reveal that the proposed approach outperforms the commercial ones for the problem of up to 100 vessels. Drones have also been well utilized for military purposes [13–16]. For instance, the routing of a set of drones to destroy a determined group of targets that are prioritized differently is studied in [13]. The authors propose a two-phase approach that considers resolving a target assignment subproblem for each drone in the first phase. The second phase in this solution framework is to solve a travelling salesman problem (TSP) to obtain a routing plan. Similarly, the authors in [14] develop an integer-programming model for an environment where new targets may emerge dynamically. This model reassigns UAVs to the updated group of tasks regarding any changes in the battleground. In [15], the routing of drones is considered for military surveillance purposes, where drones gather information from targeted area using sensors. The proposed strategy chooses the sensors for each drone by including payload capacity restrictions. Then, following these constraints, a group of drones is routed to develop a region-sharing approach by considering uncertainty on the data gained from observations. This strategy dynamically sends drones to gather information instead of focusing on a predesignated routing plan. The results of the study prove that the proposed approach is effective in a contemporary battleground where communications between drones and ground stations are frequently blocked.

Besides military purposes, several studies focus on routing drones in logistic delivery operations [17–19]. In this regard, drones assist trucks to deliver items to the customers who are located geographically. However, drones are often limited to carry only one package that makes the routing decision of the drones easier and enhance the operational efficiency [12]. For instance, a joint scheduling problem for trucks and drones is studied in [27]. In this study, drones are used to deliver packages to customers close to the storage, and trucks are responsible to deliver parcels far. The results reveal that, with such a delivery system, customers receive their orders faster. Moreover, it reduces cost of distributions as well as environmental impacts. Another similar problem is studied in [18]. In this study, trucks are permitted to carry drones in specific routes so that the drones can fly from the trucks and deliver parcels to the customers who are far from the storage. In [19], the authors prove that the potential improvement in delivery efficiency of the cooperation between drones and trucks depend on the speed of drones and the square root of the ratio of the speed of trucks.

UAVs or drones are increasingly proposed for medical use cases as well. For example, the study in [20] develops a new optimization model to help in the deployment of a network of automated external defibrillator (AED)-enabled medical drones to minimize the time it takes to reach to a patient's side. The proposed approach can optimally locate drones by considering the problem of backup coverage location with complementary coverage. At the same time, it improves backup coverage with insignificant loss of initial coverage.

In several studies, drones have been used for the tasks related to trajectory planning and task allocation [21,22]. For example, in [21], the authors propose an automated surveillance system to track several mobile ground targets. The aim of this study is to reduce the total energy consumption and to find the exact location of the targets. The study in [22] proposes a system containing several operating drones and a control station. The drones receive control information from the control station and send their location information and the sensed parameters back to the control station. The results show the effectiveness of the proposed task allocation algorithm in terms of task completion.

Apart from the studies where drones are used for ideal trajectory planning problems, some studies have utilized drones to track different targets using various sensors. For example, in [23], an algorithm is proposed to track a mobile target in a cooperative manner using several drones equipped with cameras. The goal is to keep the mobile target in the position visible by cameras from various angles while achieving a low computational complexity. In addition, the authors in [24] investigated a similar problem by considering multiple criteria, such as the number of drones, the satisfaction of customers, and the total distance moved by the drones simultaneously. The objective is to detect the exact location of mobile targets using the sensors placed on the target.

Optimal drone placement (ODP) problem and cost-effective target coverage of drones are extensively investigated in several studies [25–27]. For example, the optimal placement of a group of drones is considered in [26], with the assumption that a large number of drones are available to cover a group of mobile targets. The main objective of this study is to reduce the total amount of energy consumption. A similar study is presented in [27], where mobile targets are monitored by a group of drones that have restricted energy resources. If energy of a drone is depleted, then another drone can replace it. The results of the study reveal promising performance showing an acceptable trade-off between the computational effort and the quality of the solution. In [25], the authors propose a mathematical model to formulate the ODP problem. They provide an improved model that considers the energy of each drone, and design an ideal approach to solve the placement problem of static or mobile drones. Using two low-complexity centralized algorithms, samples of the mentioned problem with more than 50 targets and a large number of possible locations for the drones can be solved.

8.3 Proposed Methodology

One of the major drawbacks of drones is their limited range [28], which is due to the capacity of the battery. Another challenging issue regarding the usage of drones is related to their high price. Therefore, there is a need to optimize the number and location of drones to have full coverage of an area. In this regard, we propose a method called optimized placement approach (OPA) to minimize the total number of drones required while providing maximum coverage. This in turn leads to reduction in cost.

8.3.1 Optimized Placement Approach

Let D denotes a group of available drones and T represents the group of targets to be monitored assuming that each target is determined by its coordinate (x, y, z), where x, y, and z signify length, width, and height, respectively. Therefore, given a drone D, it is located at a coordinate (x, y, z) with a target T to be monitored. It is possible to define the distance between D and T_i when $z = 0$ as follows:

$$U_{t_i}^{x_d\, y_d} = \sqrt{\left(x_{t_i} - x_d\right)^2 + \left(y_{t_i} - y_d\right)^2} \tag{8.1}$$

Drones have a visibility of θ, which is signified by a disk on the plane with radius r^z depending on z_d. The drone visibility is also dependent on the angle of camera lens. Moreover, the position where each drone $d \in D$ should be located (x_d, y_d, z_d), and the target $t_i \in T$ monitored by the drone should be decided. Therefore, the first decision variables are defined as follows:

$$\delta_{xyz}^d = \begin{cases} 1 & \text{if } d \text{ is located at coordinate } (x, y, z) \\ 0 & \text{otherwise} \end{cases} \tag{8.2}$$

$$\gamma_{t_i}^d = \begin{cases} 1 & \text{if taret } t_i \text{ is observed by drone } d \\ 0 & \text{otherwise} \end{cases} \tag{8.3}$$

The goal is to carefully watch and monitor all targets with at least one drone to minimize the number of drones required as well as total energy consumption. Furthermore, the energy consumption of each drone is formulated as follows:

$$E = (\beta + \alpha k)t + P_{max}(K/S) \tag{8.4}$$

where β is the minimum power for the drone required to stay in the air and α is a motor speed multiplier. Both α and β are dependent on the weight of the drone, and the characteristics of the motor it is using. P_{max}, S, and t are maximum power of the motor, speed, and the operating time, respectively. The term αk indicates the relation between power and height, and $P_{max}(K/S)$ is the power consumption required to move up to height K with speed S. The objective function is to minimize the total number of drones, which is formulated as follows:

$$\text{Min } f(\delta) \; s.t \sum_{(x,y,z)} \delta_{xyz}^d \leq 1 \; \forall \quad d \in D \tag{8.5}$$

$$\gamma_{t_i}^d \leq \sum_{(x,y,z)} \delta_{xyz}^d \left(\frac{r^{z_d}}{D_{t_i}^{dxy}} \right) \quad \forall \; d \in D, \; t_i \in T \tag{8.6}$$

8.3.2 Equations of the Drone's Location

The flying zone for the drone is represented by Z_{max}. Detection of the target above Z_{max} is not possible, and the drones are not permitted to fly above this threshold in the region. Drones also cannot fly below Z_{min}. The flying zone is presented by a rectangle of length X_{max} and width Y_{max} such that

$$\sum_{d \in D} \gamma_{t_i}^d \geq 1 \quad \forall t_i \in T \tag{8.7}$$

When the drones fly, they need to observe the target for a specified amount of time. Additionally, the target can move in the region, particularly a time window $\left[\tau_{min}^{t_i}, \tau_{max}^{t_i}\right]$ is associated with each target $t_i \in T$, meaning that at the beginning the target t_i is placed at the point of coordinate, and it has been detected in the time range specified by the time window.

If the target is moving to catch mobility in the system, a sequence coordinate C_i is associated with each target:

$$|C_i| = \left[\frac{\tau_{max}^{t_i} - \tau_{min}^{t_i}}{\Delta\tau} \right] \tag{8.8}$$

where $\Delta\tau$ is the time interval in which a new position of the target t_i is reached.

Considering all the constraints and objectives discussed earlier, the equations for minimizing the number of drones and the total energy consumption can be formulated as follows:

$$f(\delta) = \sum (x, y, z) \sum_{d \in D} \delta_{xyz}^d t \tag{8.9}$$

$$E = \beta \sum (x, y, z) \sum_{d \in D} \delta_{xyz}^d t + \alpha \sum (x, y, z) \sum_{d \in D} Z\delta_{xyz}^d$$

$$+ \frac{p_{max}}{s} \sum (x, y, z) \sum_{d \in D} Z\delta_{xyz}^d \tag{8.10}$$

8.4 Performance Evaluation and Results

In this section, to assess our proposed model, we discuss the performance metric and parameters as well as the results obtained by simulation.

8.4.1 Simulation Setup

Equations (8.1)–(8.10) in Section 8.4 are aimed to minimize the number of drones. The simulation has been implemented in Octave programming language, namely GNU 4.4.1 [29] which is a high-level scientific programming language primarily intended for numerical computations. The script contains the following three files.

complete_coverage.m which performs all the calculations regarding targets and other parameters.
plot_drone_coverage.m which plots the graph that represents the drones and targets monitored by drones.
test_complete_coverage.m which contains the test case.

The script was executed on a device with Windows 8 Operating System (OS). The device has the following specifications: Intel(R) Atom(TM) CPU Z2760 @ 1.80 GHz, 1,800 MHz, 2 Core(s), 4 Logical Processor(s). The usage of RAM was low, and the computation time was from 4 to 6 s. There are a couple of assumptions that were made during the simulation phase. First, the battery capacity is not subjected to optimization, that is, the optimum value for battery capacity corresponding to the minimum number of drones are determined through trial and error. Second, each drone covers an area that is a square of 1 km^2. Finally, the communication range between drones is considered as circular disks for simplicity purposes.

8.4.2 Performance Metrics and Parameters

To assess the proposed framework, the following performance metric is considered in the script.

Target Coverage: This is the coverage area of drones while flying over a target. The metric is evaluated while varying the following parameters:

Energy (E): It represents the initial capacity of the drone's battery.

Visibility angle (θ): It is the opening of the drone visibility range.

Horizontal Energy Consumption (γ): This is the energy consumed due to the horizontal movement of the drone.

Vertical Energy Consumption (α): It represents the energy consumed due to the vertical movement of the drone.

Number of Targets (nt): It represents the number of targets to be monitored by the drones.

8.4.3 Results and Discussions

In this section, we discuss the results obtained from the simulation tool implemented to evaluate the performance of the proposed framework. During the simulation, we considered two phases: static and dynamic targets.

8.4.3.1 Static Targets

Static targets have fixed positions and do not change their locations. We considered two situations regarding static targets. First, given a fixed number of targets ($nt = 100$), the results for the optimal solution (e.g. minimum number of drones) are obtained as follows.

Figures 8.1–8.5 show the target coverage by drones with respect to the change of energy. Figure 8.1 shows a scenario where the drones, which are flying away either horizontally or vertically or both, have a narrower visibility range than the ones close to the (0, 0) coordinates. The number of drones is initially 23 drones. If we decrease the energy consumption of flying horizontally to $\gamma = 2$ with less battery capacity as in Figure 8.2, it can be seen that drones flying to the right and to the

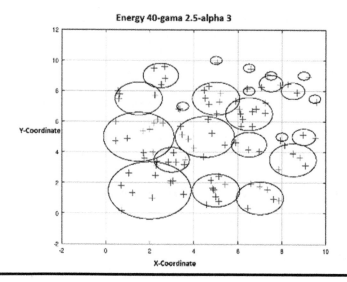

Figure 8.1 Energy = 40, gamma = 2.5, alpha = 3.

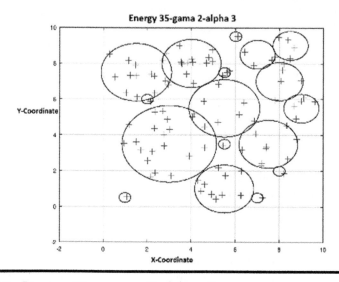

Figure 8.2 Energy = 35, gamma = 2, alpha = 3.

top right of Figure 8.2 are covering more targets due to flying further. Doing the same steps of decreasing energy cost of moving horizontally and the battery capacity makes the drones fly further and higher. Therefore, a bigger visibility range is achieved as shown in Figure 8.3.

More optimal results can be reached by increasing the battery capacity while maintaining the cost of flying horizontally and vertically. Figure 8.4 shows the least

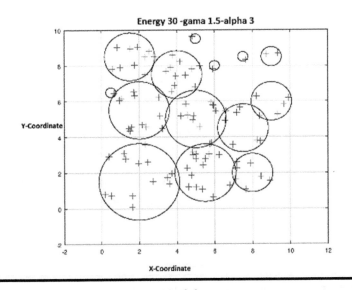

Figure 8.3   Energy = 30, gamma = 1.5, alpha = 3.

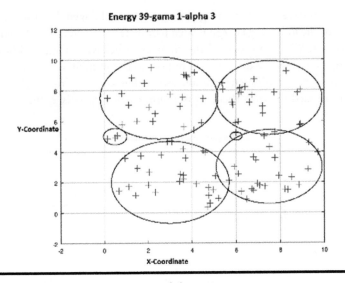

Figure 8.4   Energy = 39, gamma = 1, alpha = 3.

number of drones with wider visibility range than the ones on Figure 8.5, where the drones' visibility range is smaller. Drones that are flying further away from the hub (zero coordinates) have less visibility range by the time they reach their destination. This is because traveling further diagonally consumes more energy than flying close to the hubs (due to the short distance traveled). Therefore, this leads them to decrease their elevation and their visibilities simultaneously.

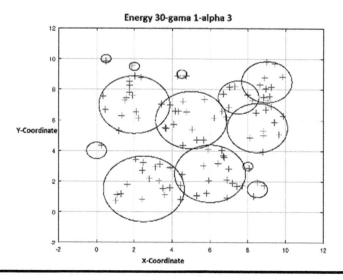

Figure 8.5 Energy = 30, gamma = 1, alpha = 3.

In the second scenario regarding static targets, we changed the number of targets (*nt*) and kept the following energy-related parameters constant; $E = 40$, gamma = 2, and alpha = 4. We then observed the changes in behavior. Figures 8.6–8.10 represent target coverage by drones with respect to the number of drones. Figure 8.6 shows three targets in different locations, since they are away from each other; three drones are needed to cover them. If the number of targets is *x*, then the number

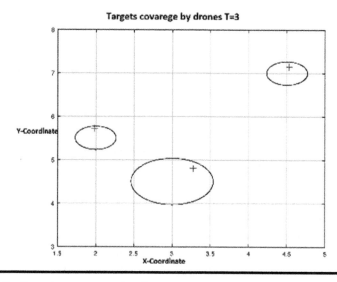

Figure 8.6 Targets coverage *nt* = 3.

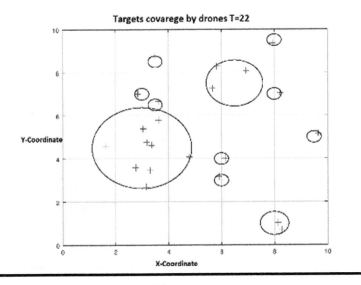

Figure 8.7   Targets coverage *nt* = 22.

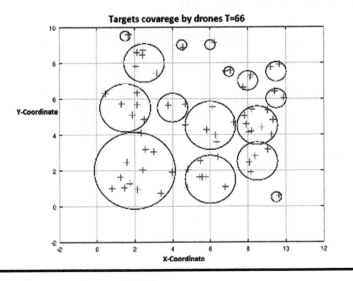

Figure 8.8   Targets coverage *nt* = 66.

of drones can have a value ranging between a minimum of 1 and a maximum of *x*. As the number of targets increases, the number of drones increases as seen in Figures 8.7–8.10.

Figure 8.11 shows the number of drones with respect to the number of targets. As it is expected, the number of drones needed to monitor a given number of targets increases gradually. However, this is only true, until a certain number

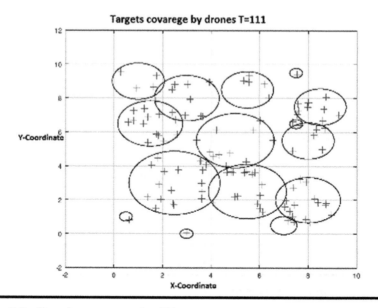

Figure 8.9   Targets coverage *nt* = 111.

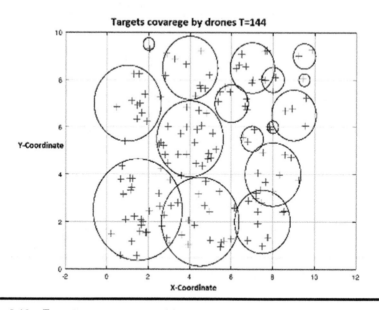

Figure 8.10   Targets coverage *nt* = 144.

of targets. This number of targets requires the highest number of drones to be monitored. It can be thought to cover several varying regions across the total area being monitored. Later, if the number of targets is further increased, the number of drones does not increase.

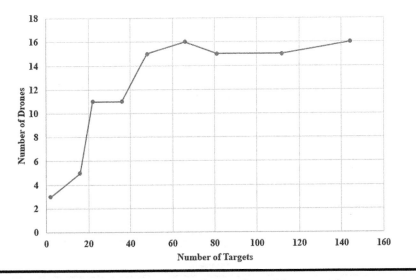

Figure 8.11   The relationship between the number of drones and targets.

8.4.3.2 Dynamic Targets

So far, we have assumed that the targets are static (not moving). However, in this scenario, we assume that the targets are capable of moving throughout the time they are being monitored. To display this effect in the simulation, the *seqLength* (S_L) and *walkArea* (W_A) parameters change. As explained earlier, S_L is the number of time intervals in which the target moves (i.e. how many times the target will move), and W_A is how far the target moves within every time interval. For example, if $S_L = 2$ and $W_A = 3$, then each target will move randomly twice, each with three steps. We set the energy-related parameters as $E = 40$, gamma $= 2$, and alpha $= 4$ to observe the changes. The script is written in such a way that if the target is static it draws a plus (+) sign, otherwise, it draws a minus (−) sign, and displays the trajectories of movement of the drones. Figure 8.12 shows static and non-moving targets, whereas in Figure 8.13, the targets are moving slightly. The visibility range of drones is slightly overlapping. This is due to the fact that the drones are trying to cover the moving targets.

In Figure 8.14, each target moved five times, each time with five steps, where each colored segment shows the trajectory of a certain target. As it can be seen in Figure 8.14, the targets are moving a lot more than the previous ones. Therefore, the overlapping region between the drones' visibility range is higher. The reason for this overlapping is the long trajectories of the targets, where they get too close to each other.

Observing the change in the target's behavior and the number of targets with respect to S_L and W_A parameters, we notice that the longer path the trajectory target takes, the closer the targets get to each other. Therefore, less number of drones is

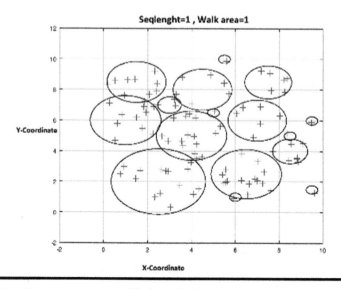

Figure 8.12   Targets coverage with $S_L = 1$, $W_A = 1$.

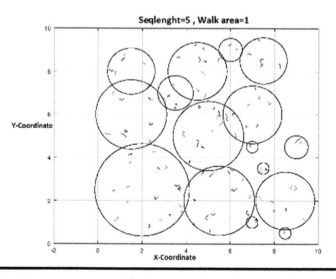

Figure 8.13   Targets coverage with $S_L = 5$, $W_A = 1$.

needed to cover them. Figure 8.15 shows how the number of drones decreases as the trajectory length increases. The trajectory length (T_J) was calculated using the following equation:

$$T_J = W_A * S_L \qquad (8.11)$$

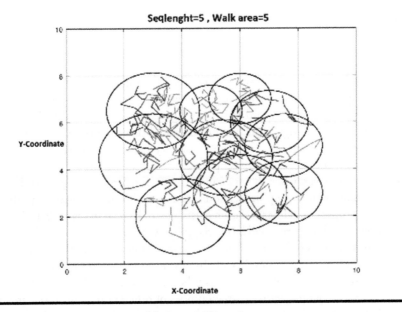

Figure 8.14   Targets coverage with $S_L = 5$, $W_A = 5$.

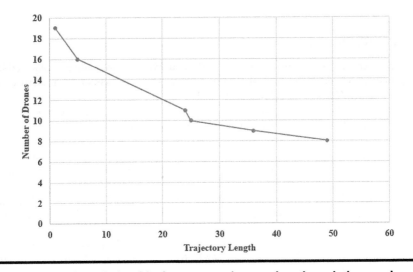

Figure 8.15   The relationship between trajectory length and the number of drones.

As it can be seen in Figure 8.15, the curve clearly displays the inverse relationship between the number of drones and the corresponding lengths of the trajectories of targets. This is clearly observed as the highest number of drones corresponds to the lowest trajectory length and vice versa.

8.5 Conclusion

In this study, we dealt with a cost minimization problem related to the optimal placement of drones to monitor a set of static or dynamic targets. Our minimization problem aims to minimize the number of drones, given a constant value of battery capacity. The problem stated earlier is formulated, and the mathematical models were provided accordingly. The simulation results obtained from different variations in changing the parameters reveal that increasing the battery capacity leads to an increase in the drone's visibility range, and thus, a decrease in the number of drones. This effectively provides a better solution for our minimization problem. Moreover, when dynamic targets are considered, moving with higher W_A leads to targets ending up in locations close to each other. In actuality, almost an inversely proportional linear relation exists, as can be seen in Figure 8.15. Therefore, the drones' visibility areas will be overlapping, which may cause a number of drones to be considered as redundant, leading to a smaller number of drones. Finally, there exists a limit where the number of drones no longer proportionally increases in relation to the number of targets. This is because the limit exhibits a case where the targets are distributed across a large number of different regions in the area monitored, rendering a further increase to the targets that does not require an increase in the number of drones needed to monitor them.

References

1. E. Ever, F. Al-Turjman, H. Zahmatkesh, and M. Riza, Modelling green HetNets in dynamic ultra-large-scale applications: A case-study for femtocells in smart-cities, *Computer Networks*, vol. 128, 78–93, 2017.
2. Y. Liu, B. Bai, and C. Zhang, UAV image mosaic for road traffic accident scene, in *32nd IEEE Youth Academic Annual Conference of Chinese Association of Automation (YAC)*, May 2017, pp. 1048–1052.
3. J. S. Álvares, D.B. Costa, and R. R. S. D. Melo, Exploratory study of using unmanned aerial system imagery for construction site 3D mapping, *Construction Innovation*, vol. 18, 301–320, 2018.
4. M. Micheletto, V. Petrucci, R. Santos, J. Orozco, D. Mosse, S. F. Ochoa, and R. Meseguer, Flying real-time network to coordinate disaster relief activities in urban areas, *Sensors*, vol. 18, no. 5, 1662, 2018.
5. L. Zhu, F. R. Yu, Y. Wang, B. Ning, and T. Tang, Big data analytics in intelligent transportation systems: A survey, *IEEE Transactions on Intelligent Transportation Systems*, vol. 20, 383–398, 2018.
6. R. Tariq, M. Rahim, N. Aslam, N. Bawany, and U. Faseeha, DronAID: A smart human detection drone for rescue, in *15th IEEE International Conference on Smart Cities: Improving Quality of Life Using ICT & IoT (HONET-ICT)*, October 2018, pp. 33–37.
7. A. Gao, S. Wu, F. Wang, X. Wu, P. Xu, L. Yu, and S. Zhu, A newly developed unmanned aerial vehicle (UAV) imagery based technology for field measurement of water level, *Water*, vol. 11, no. 1, 124, 2019.

8. V. Sharma, G. Choudhary, Y. Ko, and I. You, Behavior and vulnerability assessment of drones-enabled industrial internet of things (IIoT), *IEEE Access*, vol. 6, 43368–43383, 2018.

9. J. C. Padró, V. Carabassa, J. Balagué, L. Brotons, J. M. Alcañiz, and X. Pons, Monitoring opencast mine restorations using Unmanned Aerial System (UAS) imagery, *Science of the Total Environment*, vol. 657, 1602–1614, 2019.

10. N. Al Amir, A. Marar, and M. Saeed, Eye in the Sky: How the Rise of Drones will Transfrom the Oil & Gas Industry, in *Abu Dhabi International Petroleum Exhibition & Conference*. Society of Petroleum Engineers, November 2018.

11. J. Seo, L. Duque, and J. Wacker, Drone-enabled bridge inspection methodology and application, *Automation in Construction*, vol. 94, 112–126, 2018.

12. J. Xia, K. Wang, and S. Wang, Drone scheduling to monitor vessels in emission control areas, *Transportation Research Part B: Methodological*, vol. 119, 174–196, 2019.

13. V. K. Shetty, M. Sudit, and R. Nagi, Priority-based assignment and routing of a fleet of unmanned combat aerial vehicles, *Computers & Operations Research*, vol. 35, no. 6, 1813–1828, 2008.

14. C. C. Murray, and M. H. Karwan, An extensible modeling framework for dynamic reassignment and rerouting in cooperative airborne operations, *Naval Research Logistics (NRL)*, vol. 57 no. 7, 634–652, 2010.

15. F. Mufalli, R. Batta, and R. Nagi, Simultaneous sensor selection and routing of unmanned aerial vehicles for complex mission plans, *Computers & Operations Research*, vol. 39, no. 11, 2787–2799, 2012.

16. Y. Xia, R. Batta, and R. Nagi, Controlling a fleet of unmanned aerial vehicles to collect uncertain information in a threat environment, *Operations Research*, vol. 65, no. 3, 674–692, 2017.

17. C. C. Murray, and A. G. Chu, The flying sidekick traveling salesman problem: Optimization of drone-assisted parcel delivery, *Transportation Research Part C: Emerging Technologies*, vol. 54, 86–109, 2015.

18. X. Wang, S. Poikonen, and B. Golden, The vehicle routing problem with drones: Several worst-case results, *Optimization Letters*, vol. 11, no. 4, 679–697, 2017.

19. J. G. Carlsson, and S. Song, Coordinated logistics with a truck and a drone, *Management Science*, vol. 64, no. 9, 4052–4069, 2017.

20. A. Pulver, and R. Wei, Optimizing the spatial location of medical drones, *Applied Geography*, vol. 90, 9–16, 2018.

21. W. Chung, V. Crespi, G. Cybenko, and A. Jordan, Distributed sensing and UAV scheduling for surveillance and tracking of unidentifiable targets, in *Sensors, and Command, Control, Communications, and Intelligence (C3I) Technologies for Homeland Security and Homeland Defense IV*, International Society for Optics and Photonics, vol. 5778, May 2005, pp. 226–236.

22. S. Simi, R. Kurup, and S. Rao, Distributed task allocation and coordination scheme for a multi-UAV sensor network, in *Tenth IEEE International Conference on Wireless and Optical Communications Networks (WOCN)*, July 2013, pp. 1–5.

23. J. Kim, and Y. Kim, Moving ground target tracking in dense obstacle areas using UAVs, *IFAC Proceedings Volumes*, vol. 41, no. 2, 8552–8557, 2008.

24. F. Guerriero, R. Surace, V. Loscri, and E. Natalizio, A multi-objective approach for unmanned aerial vehicle routing problem with soft time windows constraints, *Applied Mathematical Modelling*, vol. 38, no. 3, 839–852, 2014.

25. D. Zorbas, L. D. P. Pugliese, T. Razafindralambo, and F. Guerriero, Optimal drone placement and cost-efficient target coverage, *Journal of Network and Computer Applications*, vol. 75, 16–31, 2016.
26. D. Zorbas, T. Razafindralambo, and F. Guerriero, Energy efficient mobile target tracking using flying drones, *Procedia Computer Science*, vol. 19, 80–87, 2013.
27. L. D. P. Pugliese, F. Guerriero, D. Zorbas, and T. Razafindralambo, Modelling the mobile target covering problem using flying drones, *Optimization Letters*, vol. 10, no. 5, 1021–1052, 2016.
28. E. Yanmaz, S. Yahyanejad, B. Rinner, H. Hellwagner, and C. Bettstetter, Drone networks: Communications, coordination, and sensing, *Ad Hoc Networks*, vol. 68, 1–15, 2018.
29. GNU Octave, Available at: https://gnu.org/software/octave/.

Chapter 9

Price-Based Data Routing in Dynamic IoT

Fadi Al-Turjman
Antalya Bilim University

9.1 Introduction

Internet of things (IoT) is a pervasive technology for applications ranging from smart grid to vehicular networking and smart homes to smart workplace. IoT is growing as a framework to encompass all identifiable things, in a dynamic and interacting network. The promise of clever approaches and dynamic systems that could benefit from the aggregation and analysis of information over the IoT infrastructure is quite pervasive. Scientists in networking, research and development (R&D) divisions, and many businesses are in the race to develop an achievable and robust architecture to realize the IoT paradigm [1,2].

Yet, many hindrances render the IoT framework mostly a challenge. To date, much has been proposed on the promise and benefits of IoT, yet far less has covered the routing protocols to actually operate such a dynamic and large-scale paradigm [3,4]. The vision, however sparse, promises a robust and dynamic framework to integrate many enablers that are already outshined in research and development.

Obviously, wireless sensor networks (WSNs) are envisioned to play a dominant role in IoT paradigms. The resilience, autonomous, and energy-efficient traits of WSNs render them a vital candidate for dominating the information collection task of an IoT framework [5,6]. Equally vital, the use of RFID (radio frequency identification) technologies for non-LOS (line of sight) and seamless identification of objects is gaining much prominence as a key player in IoT frameworks [7]. The low cost associated with deploying RFID tags (passive or active) is an important motivation. In fact, some argue that RFIDs have been a main motivator for the IoT framework [7,8].

The integration of these enablers, along with internet-based and context-aware services, facilitates a dynamic platform for IoT. Nevertheless, much of current research has focused on developing these enablers in segregation and optimizing their performance under local constraints and objectives. One of the most important tasks to be carried out, in such a large-scale and dynamic environment, is relaying information from a source to a destination, given the new emerging characteristics in IoT. Typical routing approaches consider that all components belong to the same owner/provider; hence, routing costs and link weights are directly proportional to their local provider characteristics. Though, IoT routing becomes inherently complex by multiple factors. An intrinsic design factor in IoT is delay tolerance [9].

In reality, an IoT node has only partial knowledge regarding the full path to the destinations assigned to the packets it delivers. Due to splitting, which is mainly caused by nodal mobility, connectivity may occur on an irregular basis. In such circumstances, nodes are required to store and carry data packets until an appropriate forwarding chance ascends in a store-carry-forward fashion [10]. Typical routing approaches of sensor networks are unfortunately mobility-intolerant, since most of the WSN network architectures assume stationary sensor nodes [11,12]. As stated earlier, we adopt an expanded notion of sensor networks that incorporates MANet (mobile ad hoc network) nodes. An abundance of routing-layer protocols has been proposed to accommodate the dynamic topology in MANets and WSNs [11,12].

Yet, for all these protocols, it is implicitly assumed that the network is connected, and there is a contemporaneous end-to-end path between any source/destination pair. In other words, the topology in the standard dynamic routing problem is assumed to be always connected, and the objective of the routing algorithm, hence, is confined to finding the best currently available full path to move traffic from one end to the other. Unfortunately, none of these assumptions stand in a delay-tolerant setup. An IoT data delivery scheme must be delay tolerant to cope with intermittent connectivity, in addition to providing faster delivery alternatives for other delay-sensitive types of data that demand minimal delays.

Furthermore, most entities participating in sensing, identification, and relaying in IoT belong to different networks with multiple owners. It is not in the best interest of such networks to allow its resources to be utilized for relaying data across the network, without compensation. For example, an intermediate relay node, belonging to a WSN for surveillance, would not freely take part in relaying information of nearby RFID readers or other WSNs. Thus, price and trading, in addition to all of the routing metrics that govern a mesh ad hoc network, need to be considered before a suitable routing protocol is presented to relay packets across an inherently diverse IoT.

To this end, we define an *IoT Setting* by the following four main characteristics: (1) cost-effectiveness, (2) seamless integration, (3) reliability and trust, and

(4) delay tolerance. Hence, we provide a framework, encompassing a cost-efficient IoT architecture, to address data delivery objectives according to the aforementioned characteristics of IoT setting. The design objectives are to be met with respect to metrics such as delay tolerance, cost, and power saving. Our proposed framework makes use of ubiquitous relays available in today's topologies to enhance connectivity and delivery rates between the components of the integrated topology. Our framework will as well provide delivery guarantees with respect to delay and connectivity over end-to-end links. Such guarantees will be carried out by dedicated components of our integrated architecture, in addition to other components incorporated within the wider IoT vision.

Our impact in this work is twofold. First, presenting a routing approach customized for heterogeneous IoT components. This is only possible with our second contribution, a pricing model that caters for the diverse requirements and conditions of nodes that are willing to relay IoT data packets without using the internet backbone. This pricing model presented here joins measures of load balancing, delay, buffer space, and link maintenance. An outline of the targeted routing problem and the dynamic constituents of the envisioned IoT is depicted in Figure 9.1.

The rest of this chapter is organized as follows. Section 9.2 covers the background on IoT routing and its enabling technologies. Then, a rigorous definition of our proposed network model manifesting the interactions of components in the IoT and their governing constraints is proposed in Section 9.3. Next, we formally present our adaptive routing protocol in Section 9.4. Our proposed model is verified in Section 9.5 via use cases and Markov chain in Q-theory. Extensive results are performed and described in Section 9.6. Finally, our work is concluded in Section 9.7.

9.2 Background

Nowadays, everything around us is surrounded with different types of networks. Wireless Fidelity (WiFi), long-term evolution (LTE) wireless communications, broadcasting, streaming, etc. are quite common widely spread technologies. However, they bring their own limitations. These limitations can be in the form of cost or technology. Most often, it is about the cost of maintaining and placing an efficient network that can integrate all for what we call the IoT. Several attempts have been made for improvements and performance gains in the enablers of IoT (especially WSNs and RFIDs). To present a perspective on these enablers and the major domains of properties, Table 9.1 summarizes three main paradigms to IoT. Accordingly, we emphasize two major driving forces. First, the lack of a distinctive routing approach that caters for dynamic IoT. The second drive lies in the trade-off costs of routing over multiple entities, belonging to different service providers.

A major misconception was imposed by an inherent property of IoT, namely being a descendent of the internet. That is, as research on IoT developed, it was

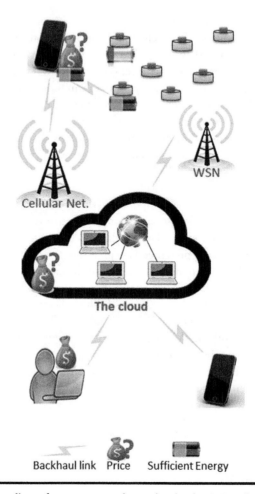

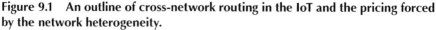

Backhaul link   Price   Sufficient Energy

Figure 9.1   An outline of cross-network routing in the IoT and the pricing forced by the network heterogeneity.

expected that a significant pool of protocols previously developed for internet services would migrate into the IoT. Nonetheless, as the IoT is set to encompass many stationary (static WSNs, RFID readers, etc.) and dynamic (laptops, personal digital assistants (PDAs), cell phones, etc.) components, we are challenged with multiple issues [2,13]. Most importantly, assuming that all components will intercommunicate via the internet is insufficient and often degrading the intended dynamic paradigm performance.

A major obstacle would stem from the mounting number of messages that overload a network already handling millions of hosts. This is a noteworthy problem as recent endeavors are targeting higher levels of dynamic interaction between the IoT and its users, as in the human–computer interaction work presented by

Table 9.1   IoT Enablers and Their Properties

Property	Wireless Networks			
	IoT	*MANets*	*WSNs*	*RFIDs*
Topology	Dynamic	Dynamic[a]	Mostly static	Application dependent
Buffer size	Varies	High	Low	None
Medium contention	High	High	Medium	Low with singulation
Mobility	Frequent	Varies[a]	Limited	Frequent
Communication range	Varies	High	Medium (varies)	Reader dependent
Typical density	Very high	Small to medium	Medium to high	Medium to high
Computational power per node	Varies	High	Low	Low to none[b]
Internode communication	Heterogeneous	Homogeneous		

[a] Since MANets encompass VANets as well.
[b] Disregarding active tags, as they equate many features of SNs.

Kranz et al. [14]. As such, if a WSN needs to identify an object, with the aid of an RFID reader, direct communication between a sensing node (SN) and the reader would influence bottlenecks of communication and swarming the backhaul over the internet with numerous packets. This is a prominent architecture, one that is strongly pushed for as a truly integrating IoT [15]. There is a need for establishing a cooperative scheme for routing in the IoT: one which includes all nodes with capabilities of relaying data. This includes those with only one access medium (e.g. WiFi routers) and others with multiple mediums (e.g. cell phones). Yet, due to obvious reasons of resource conservation, such entities would not participate in relaying data packets unless there is an incentive [16]. It is vital to note that some components only generate data (e.g., IDs), such as RFID tags.

Different incentives take part in the pricing model that dictates the choice of a group of candidates for relaying. Recent results in incentive-based routing have been well studied. Zhong et al. [17] present an elaborate study on routing and forwarding in MANets, by emphasizing a scheme that ensures optimal gain for the individual nodes. Auction pricing patterns [18] allocate resources to users through a bidding process conducted by the users. Auction pricing can equally accomplish resource allocation and service attributes. The scheme is based on profile bids, where the seller computes an allocation to be given to the buyer. This sequence is repeated until all parts agree. We note that the lengthy negotiation process is ineffective, particularly, for mobile users while in high-speed transit.

Dynamic priority pricing schemes [19,20], on the other hand, are applied on a wireless link shared by the subscribers, divided into different priority classes by the service provider. The mobile subscriber is allowed to select the preferable transmission rate and its traffic allocation. In addition, the subscribers can split their traffic among several priority classes and be charged accordingly. The efficiency of this scheme is in its simplicity and scalability. The provider's profit increases according to the user's satisfaction by the service. Priority schemes assume, however, that the network's capacity buffer is not exceeded and priority thresholds are kept under a maximum level. Other schemes have been presented to incorporate dynamic game theory models, for noncooperative scenarios where local utility functions dictate the participation of nodes in relaying [21,22]. It is important to note as well that many of such factors are nontrivial to compute, and many nodes in the IoT would not possess the computational capacity to compute and execute local utility functions. Thus, it is intuitive to pursue a game-theoretic approach for the IoT only if it caters for offloading the task of computing local utility functions to nearby high-end nodes.

Other problems stem from scalability issues in IoT, being an architecture that is envisioned to span continents and vast distances [23]. The major issue is being able to establish and maintain end-to-end links and keeping track of nodes that are dynamically entering and exiting from the network. Remedies have been proposed by increasing the density of backhaul connections and multiple readers to enhance connectivity and capacity, respectively. However, recent studies highlighted the degrading effect of inter-reader and relay collisions [24].

9.3 IoT System Model

Many factors are intrinsically dominant in the operation of a routing protocol. More factors are further augmented as we devise a routing protocol for the IoT paradigm with dynamic topologies and heterogeneous data generating/sharing systems in place. In such a comprehensive paradigm, an incentive data sharing policy is required to motivate sensor owners to participate in the sensing process and to ensure that the provided data is fairly priced. And this in turn necessitates addressing IoT-specific challenges such as system's limitations in terms of lifetime, available capacity, reachability, and delay. In addition, a careful focus on quality management and assurance constraints is to be considered as well.

Thus, it is the scope of this section to detail and elaborate upon the factors that are considered in IoT-specific routing protocols that tackles all the aforementioned concerns. No single protocol would achieve all objectives, as many objectives are inherently contradictory, thus routing belongs to the notorious no free lunch class of algorithms.

Our system is presented in the remainder of this section and elaborated upon in four components. First, we present the IoT network as a whole. In the following subsection, we discuss each of the resources pertaining to these nodes and affecting the relaying scheme. The discussion is complete with a derivation for the utility functions that would govern the choice of nodes. In Table 9.2, a summary table of used notations is presented.

Table 9.2　Summary of Notations

Notation	Description
N	Number of in-network devices.
n_i	A node/device $i \in N$.
δ	A threshold on number of hops per routed packet.
Ψ_i	A quintuple computed for each $n_i \in N$ based on residual energy, delay, trust, and capacity per buffer.
u_i	Available storage capacity to compute and relay a message at n_i.
u_i'	Normalized buffer capacity per n_i.
π_i	Power consumption per n_i.
π_i'	Normalized power consumption per n_i.
E_i	Maximum charge per node n_i. It varies from n_i node to another in a heterogeneous IoT.
D_k	Kth data packet size in the queue.
E_{ij}	Euclidian distance between a source node i and a destination j.
ω	A delay step; the distance a wireless signal would travel in one time unit.
D_{single}	A single hop delay a packet will experience.
D_{total}	The total end-to-end delay a packet will experience.
T_{D_j}	A normalized value representing trust level of the exchanged packets between a node j and the destination D.
P_r	The probability to be connected within r communication range.
γ	Path loss exponent in a specific environment.

(Continued)

Table 9.2 (*Continued*) **Summary of Notations**

Notation	Description
μ	A normally distributed random variable with zero mean and variance σ^2.
K_0	A constant value calculated based on the mean heights of the transmitter and receiver.
λ_d, λ_t	The arrival rates for data and trusted packets, respectively.
μ_d, μ_t	The departure rates for data and trusted packets, respectively.
μ_{cd}	The rate of departures caused by finding better price in the system.
MQL	Mean queue length in the system.
RT	Response time of the system.
P_{ij}	The probability of being in an *ij* state shown in Figure 9.3b.
γ	The average percentage of transmitted packets that succeed in reaching the destination.
β	The inflection point of the randomly generated packets sequences.
ϵ	Represents the tolerance to variation in data quality expressed in a sigmoid function according to Eq. (9.5).
α	A constant that determines the rate of decrease of the utility function in Eq. (9.5).

9.3.1 IoT Model

We assume a network of heterogeneous devices, those belonging to WSNs, MANets, RFIDs, and stationary/mobile devices. Each communicating entity of these devices (i.e. wired/wirelessly enabled device) is considered as an active node in this design; hereon referred to as a node. Thus, given a set N covering all these devices, we represent each node as $n_i \in N$, where $i = \{1, 2, \ldots, |N|\}$. Thus, the set N includes both nodes that are sole relays (access points, routers, WSN sinks, etc.) and other devices with relaying capabilities (communication and processing). We assume each n_i is connected to the network, as disconnected nodes would not take part in this scheme, i.e. if there's no link from a node n_j to some other node $n_i \in N$ then $n_j \notin N$. It is important to note that the size of N varies over time as nodes enter, leave, and run out of energy.

Connectivity between nodes is assumed to take one of two nodes. If nodes are in close proximity, then we advocate for direct communication between the nodes without rerouting through the internet (via a backhaul). However, to sustain the important large-scale aspect of the envisioned IoT, we dictate that packets traveling over a threshold of hops δ would be routed through a backhaul as an intermediate stage and then rerouted to the final destination from the closest backhaul to that destination. It is thus an important factor to cater to both short- and long-range communication between nodes, both directly or via the internet backbone. We remark again the importance of obeying to approaches that utilize the internet backbone only when necessary, and reroute spatially correlated data packets between neighboring nodes without loading the backbone.

9.3.2 IoT Node

Each node $n_i \in N$ takes part in relaying as well as other tasks. Accordingly, each n_i encompasses a group of resources, with a minimum of communication and processing units. Moreover, in the case of cell phones, PDAs, WSN sinks, and RFID readers, they would all encompass a larger pool of resources, not necessarily geared towards the routing task. Thus, it is important to consider how the load of performing these tasks could affect/hinder the relaying capabilities of such nodes. We note their existence, but in this scope, we account for their effect on residual energy and buffer capacity. And thus, main design aspects considered in the utility function of an IoT-specific node is depicted in Figure 9.2a. A quintuple Ψ_i is computed for each $n_i \in N$ aggregating the following parameters, for both direct and implied effects on the routing scheme:

9.3.2.1 Residual Energy and Power Model

Each node operating on battery power would possess an energy reservoir denoted by e_i, where $0 \leq e_i \leq E_i$. Here we denote E_i as the maximum charge for n_i, since this varies across the different types of nodes. To normalize this representation across the heterogeneous nodes in this protocol, we define

$$e_i' = \frac{e_i}{E_i} \tag{9.1}$$

Knowing the size of data packet D_k to be forwarded, its distance to its next hop and the current load (u_i), each node would compute a value for the power consumption to be incurred by processing a given packet. The power consumption would be represented as π_i. However, since this is a crude number dependent on the available resources at node n_i and their strength (of transceiver), this value is normalized by dividing by its maximal attainable load and transmission distance. This would

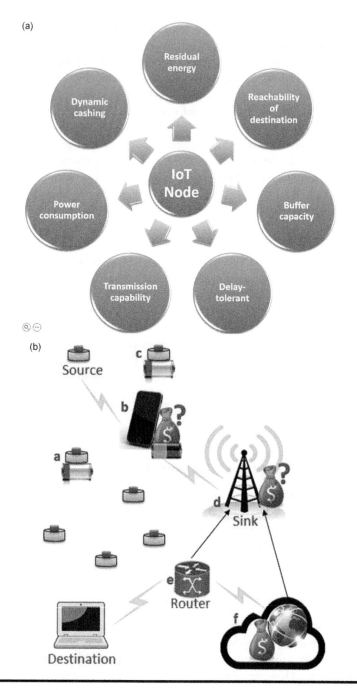

Figure 9.2 (a) The IoT-specific aspects incorporated in the pricing model for computing a utility function. (b) Use case demonstrating the routes taken by INs from the source to a destination (e.g. remote laptop).

favor high-end nodes with longer transmission capabilities and more buffers. The normalized value is represented as π'_i.

9.3.2.2 Load and Buffer Space

Since an intermediate node (INs) might be taking part in multiple tasks, each node will represent its available capacity to compute and relay a message as a utilization factor u_i, which will be normalized by opposing it to its maximal capacity, thus yielding a normalized u'_i. This is directly derived from memory and processing operations and the yield of the node's micro controlling unit (MCU) in handling different paths. Our delivery approach adopts a data delivery approach, where a network IN i has a limited capacity for the maximum amount of data that can be relayed over a specific time period. We define a normalized relaying capacity for the set of i's as

$$u'_i = \frac{u_i}{\max\_u_i} \tag{9.2}$$

where $\max\_u_i$ is the maximum expected capacity.

9.3.2.3 Delay

We define a delay step ω which is the distance a wireless signal would travel in one time unit. Let E_{ij} be the Euclidian distance between a source node i and a destination node j, then the discrete propagation delay over a single-hop link (i, j) would be $\frac{E_{ij}}{\omega}$. Hence, the discrete delay over a multihop path is the sum of discrete delays of single-hop links that constitute that path. For the sake of generality, single-hop delay (D_{single}) and total delay (D_{total}) can be defined in Eqs, (9.3) and (9.4), as follows:

$$D_{single} = \frac{E_{ij}}{\omega} + \psi \tag{9.3}$$

and

$$D_{total} = \sum_{total\ hops} D_{Single} \tag{9.4}$$

9.3.2.4 Trust

Trust parameter is a history-based function that is calculated at the network IN per destination to represent a D_j fulfillment measure. A higher T_{D_j} indicates that previous data exchanges between $node_i$ and D_j have been fulfilled according to the predefined IoT characteristics (e.g. capacity, delay, trust, etc.) promised by D_j. We also remark that this delay parameter can be defined alternatively according to the applied IoT application with varying weighting factors. For example, delay is a key parameter in risk management IoT applications. However, it can be more relaxed in other kinds of entertainment applications.

9.3.3 Pricing Model

All the previous factors are pertaining to nodal resources and their operation levels, in contrast to the remaining energy each node could support. However, an important aspect to cater for, and possibly arbitrate upon, is the price the nodes are going to charge for relaying a given data packet. That is, since the heterogeneous nodes in the IoT system neither belong to the same network nor the same owner, it is imperative that a monetary cost would be associated with the forwarding action. This is an important aspect for integrating multiple heterogeneous nodes in the architecture and enhancing global scalability. The argument for utilizing current resources with a given cost/price is more dominant than claims of deploying enough resources to cater for all connectivity and coverage tasks of the envisioned IoT. We hereby adopt and build upon the former argument in an IoT-specific pricing model.

Pricing schemes in heterogeneous networks, such as the ones in IoT paradigm, usually cover a wide range of factors to determine the value of a resource in usage. However, the most efficient schemes capitalize on the differential values of each of the heterogeneous components. We built our pricing strategy based on the original laws of supply and demand, the abundance of resources, and their homogeneity decrease of their value. And higher prices are usually assigned to nodes with rare services [25,26]. Moreover, an IoT-driven pricing model has to realize a level of service that aggregates data from several sources, including the network context (e.g. 5G, WiFi, LiFi, etc.), the mobile apps' pool, and other sources, to produce better reliable readings. We also remark that an IoT-specific pricing model shall not assume a direct provider–client relation in determining their price mechanisms [27]. In contrast, in large economic systems such as the one we are targeting in this research, entities known as INs are required for coordinating network management tasks. These INs take care of the necessary authentication, billing, and interfacing tasks to find an appropriate service provider within the heterogeneous crowd of resources in the IoT market. To this end, we define an IoT-specific setting by the following four main characteristics: (1) node residual energy, (2) load and buffer space, (3) trust level, and (4) delay tolerance. Those characteristics have been adopted into two different simulation environments: MATLAB and *Simulink* to validate our price-based results. *Simulink*, a framework built on MATLAB, is used for validation purposes to obtain more realistic results by imitating the real multi-layered networking process.

Hence, we provide a pricing framework for each node, encompassing a cost-efficient IoT architecture, to address data delivery and routing objectives according to the aforementioned characteristics of the IoT setting. We introduce γ_i, which is a pricing factor for each node in the IoT. This is a factor that could be set as a flat rate per number of bytes transmitted or computed based on the state of current resources at node n_i represented by Ψ_i. In this work, we adopt the latter as a proof of concept to the monetary exchange for forwarding the IoT under varying conditions. Thus, we denote the price charged by each node n_i as p_i:

$$p_i = \gamma_i * \left[\frac{E_{Tx}\left(D_k, n_j\right) + E_{Rx}\left(D_k\right)}{e_i} + \pi_i' + u_i' \right] \quad (9.5)$$

It is intuitive to note that owners of nodes in the vicinity of such a network may choose to adaptively contribute or withdraw from the topology by varying the value assigned to γ_i, i.e. setting it to a relatively high value would diminish the chances of it being selected for relaying.

9.3.4 Communication Model

In practice, the signal level at distance d from a transmitter varies depending on the surrounding environment. These variations are captured through the so-called log-normal shadowing model. According to this model, the signal level at distance d from a transmitter follows a log-normal distribution centered on the average power value at that point [28]. Mathematically, this can be written as

$$P_r = K_0 - 10 P_l \log(d) - \mu d \quad (9.6)$$

where d is the Euclidian distance between the transmitter and receiver, P_l is the path loss exponent calculated based on experimental data, μ is a normally distributed random variable with zero mean and variance σ^2, i.e. $\mu \sim N\left(0, \sigma^2\right)$, and K_0 is a constant calculated based on the mean heights of the transmitter and receiver.

9.4 Adaptive Routing Approach

The integrated architecture imposed by the heterogeneity of IoT demands a scalable and inclusive routing protocol. The latter property refers to the exploitation of different relaying resources that are able to carry forward a data packet towards the destination. This section presents adaptive routing approach (ARA) protocol.

ARA is divided into two stages: forward and backward. The forward stage starts at the source node by broadcasting setup messages to its neighbors. A setup message includes the cost seen from the source to the current (intermediate/destination) node. A node that receives a setup message will forward it in the same manner to its neighbors after updating the cost based on the values computed in Ψ_i. All setup messages are assumed to contain a route record that includes all node IDs used in establishing the path fragment from the source node to the current IN. The destination collects arriving setup messages within a route-select (RS) period, which is a predefined user parameter.

The backward stage starts when an Acknowledgment (Ack) message is sent backward to the source along the best selected path (called **active** path) in terms of the parameters passed in Ψ_i. If a link on the selected path breaks (due to node movement or bad channel quality), the Ack at an IN i is changed to setup message (called i_setup) and forwarded to neighbors of i which has discovered the error.

Once the source receives the i_setup, and the active path between S and D is established. When no breaks are discovered, the source receives an Ack to show that the path has been established, and it starts transmission. If during the communication session (i.e. after selecting the active path) a break is detected, the IN detecting the break will send data on an alternative route (if any) or it will buffer data and send an i_setup message to the destination to look for an alternative path.

In general, nodes can learn about their neighbors and update the routing (R) table, either by receiving a broadcasted setup message and accordingly update its neighborhood table, or by broadcasting a "*hello*" message periodically if no messages have been exchanged. This hello message is sent only to the neighborhood of the node. A new neighbor, or failing to receive from a node for two consecutive hello periods, is an indication that the local connectivity has changed.

Algorithm 1: For Source Node S

1. **If** S has a new *data* msg and no route to D
2. **Then** forward a *setup* msg.
3. **If** S receives *D_Ack* or *i_setup* msg,
4. **Then** check local p_i and send the new *data* msg's if satisfied.
5. **If** S doesn't receive a response for a route discovery (RD) period,
6. **Then** go to line 2.
7. **If** no pkts are exchanged for *hello_interval* time units,
8. **Then** send a *hello* msg and update R and p_i.

A pseudocode describing the source node algorithm is shown in Algorithm 1. Lines 1–2 represents the beginning of the forward stage, where a request to establish an active path is initiated. Such that, if S has new packets to send and no route is known to targeted destination D, then a setup message is forwarded to all available neighbors of S. To do so, all IN nodes broadcast their identity at the deployment stage and each S node keeps a record of the next hop towards some IN. Each source node n_i has a next node record that has the following: *ID field* to recognize the next relaying IN ID, the *Geo_Loc* field to determine node geographical coordinates, and the *Number_of_hops field* that has the number of hops towards the destination D. Note that this process will construct a price-based tree for each S node; such that the tree of INs that is rooted at S and involves all price-efficient INs towards the destination D will be identified at the initialization of the network. Lines 3–4 indicate that the path has been found.

Hence, active path between S and D is updated, and the source begins transmitting new data packets. Lines 5–6 describe the case where an RD period is expired. Therefore, the source restarts the RD process by sending a new setup message. Finally, lines 7–8 indicate that S has not exchanged messages with neighbors for more than *hello_interval* time units. Thus, a hello message is sent and the R is updated accordingly.

A pseudocode describing the IN algorithm is shown later. Lines 1–2 handle the forward stage, such that if an IN *i* receives a setup message, it forwards this message to all its unvisited neighbors and records every visited node to establish a backward path. Contrarily, Lines 3–7 handle the backward stage of the algorithm. If node *i* receives Ack from destination (called D_Ack), then it checks whether the neighbor towards S on the backward path is reachable or not (i.e. has a broken link). If reachable, it passes the D_Ack to this neighbor and records the necessary information to establish an active path. Otherwise, it initiates a new setup process between *i* and S by sending i_setup message to *i*'s neighbors. Lines 8–9 keep forwarding this i_setup message until it reaches S to establish an active path between *i* and S instead of the broken one.

Algorithm 2: For an IN Node *i*

1. **If** *i* receives *setup* msg,
2. **Then** check thresholds and update/forward *setup* msg if satisfied. Also, the forwarded *setup* msg records visited nodes while traveling to D.
3. **If** *i* receives *D_Ack*
4. **Then, If** a backward_neighbor is reachable,
5. **Then** forward the *D_Ack*
6. **If** backward_neighbor is not reachable,
7. **Then** send an *i_setup* msg and update R and local p_i.
8. **If** *i* receives *i_setup* msg
9. **Then** check thresholds and forward *i_setup* msg if satisfied. Also, the forwarded i_setup msg records visited nodes while traveling to destination.
10. **If** *i* receives *data* msg
11. **If** next hop is still reachable
12. **Then** send *data*
13. **If** a new active path was established
14. **Then** check the price, update R, and send *data* if satisfied.
15. **Else** buffer *data* **and** send *i_setup*

Similarly, lines 10–12 check for the availability of the next hop on the active path while data packets are transmitted through *i* towards the destination D. If next hop is not available, the IN *i* checks for an alternative path. If a new path has been established, lines 13–14 detour the data packets between S and D along this new partial route and update the active path. If no alternative path is found, line 15 buffers the data packets and initiates a new setup process. We remark that lines 2 and 9 will kill any setup message, if it is not willing to participate in routing.

Finally, a pseudocode describing the algorithm at the destination node D is shown in Algorithm 3. Lines 1–10 handle the case when a setup process has been initiated by an IN *i*. This also indicates link breakage at node *i* in active path between S and D. If there exists alternative path(s) passing through the node detecting link breakage (i.e. node *i*) or passing through the source S, lines 3–4 select the best-cost path and notify *i*. Otherwise, lines 5–10 initiate a new setup process and act as a source node in looking for a new path to S. Therefore, it sends to all D's neighbors and waits for an Ack from the source S (called S_Ack). Meanwhile, lines 11–14 represent the backward stage in response to the forward stage that has been initiated at S. The destination D keeps receiving setup messages with the corresponding found paths between S and D for an RS interval. After RS time units, D acknowledges the source S that an active path has been established by sending a D_Ack message to it through the best-cost selected path.

Algorithm 3: For the Destination Node D

1. **If** D receives *i_setup*
2. **Then** remove paths containing broken links.
3. **If** there exist path(s) passing through *i* or S
4. **Then** select best-cost path and notify *i*.
5. **If** no paths found
6. **Then** send a *setup* msg
7. **If** D receives *S_Ack* or *i_setup*
8. **Then** select path indicated by received msg.
9. **If** D doesn't receive a response for an RD period,
10. **Then** go to line 5.
11. **If** D receives *setup* msg RS not expired
12. **Then** store the candidate path and cost.
13. **If** RS expired
14. **Then** select best-cost path and send *D_Ack* on it.

9.5 Use Case and Theoretical Analysis

To demonstrate the utility of ARA protocol, we hereby adopt a use case that utilizes heterogeneous nodes in a sample IoT environment. The remainder of this use case will refer to Figure 9.2. An SN (the source) has obtained information to be sent to a destination computer. However, no direct link connects both devices, and intermediate devices belong to different networks. We assume that nodes *a*, *b*, *c*, *d*, *e*, and *f* are all willing to relay, yet *a* and *c* are already depleted in energy. The sink, node *d*, is powered by electricity and acts as an IN between the resourceful cell phone *b* and the router *e*. ARA will initiate a setup message sent to *a*, *b*, *c*, and its current neighbors. Since *a* and *c* have depleted batteries, they will terminate the flow of the setup request towards the destination. Since the cell phone *b* is in the range of communication to the source, it will forward the message to its neighbors (not highlighted here as the pattern is clear).

Eventually, the shortest path to the destination is established. The destination will receive two streams $\{S \rightarrow b \rightarrow d \rightarrow e \rightarrow D\}$ and $\{S \rightarrow b \rightarrow d \rightarrow f \rightarrow e \rightarrow D\}$. Since both *f* and *e* are resourceful entities, the arbitration of number of hops would manifest a preference for the former route, which will carry an Ack message back to the source node. It is important to note that an internet link (both forward and backward), which would also incur a cost, takes part in the route options, as the setup message would also parse through it when it is beyond the preset threshold of hops dictated by the application and source request. Furthermore, the IoT network under study can be modeled using queuing theory for steady-state evaluation, with an abstraction as illustrated in Figure 9.3a. The resulting continuous time Markov chain would be a multidimensional one for multiple types of traffic, similar to the one presented in Figure 9.3b, where λ_d and λ_t are the arrival rates, and μ_d and μ_t are

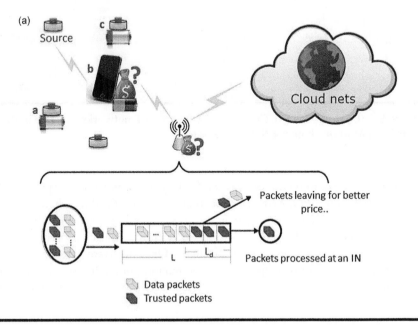

Figure 9.3 **(a) The Q-model for the IoT networks. (b) The multidimensional Markov chain for the IoT networks.**

(Continued)

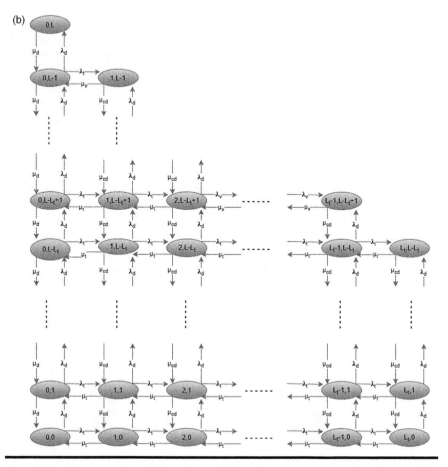

Figure 9.3 (*Continued*) **(a) The Q-model for the IoT networks. (b) The multidimensional Markov chain for the IoT networks.**

departure rates due to service completion for first-time emerging *data* packets and previously exchanged *trusted* data packets that fulfill predefined IoT characteristics, respectively. μ_{cd} is the rate for departures caused by finding better price instead of service completion in the system. Various solution methods can be employed to solve such a system for steady-state probabilities. Once the steady-state probabilities are obtained, they can be employed for computation of quality of service (QoS) measures, such as mean queue length (MQL), throughput (γ), and response time (RT) as follows:

$$MQL_t = \sum_{i=0}^{L_t}\sum_{j=0}^{L} iP_{ij} \qquad (9.7)$$

$$MQL_d = \sum_{i=0}^{L_d} \sum_{j=0}^{L} jP_{ij} \tag{9.8}$$

$$\gamma_t = \sum_{i=0}^{L_t} \sum_{j=0}^{L} \mu_t P_{ij} \tag{9.9}$$

$$\gamma_d = \sum_{i=0}^{L_d} \sum_{j=0}^{L} \mu_d P_{ij} \tag{9.10}$$

$$RT_t = \frac{MQL_t}{\gamma_t} \tag{9.11}$$

$$RT_d = \frac{MQL_d}{\gamma_d} \tag{9.12}$$

where P_{ij} is the probability of being in a state ij in Figure 9.3b. Please note that it is also possible to employ, similarly, the steady-state probabilities to compute the expected value for energy consumption of the considered price-based IoT system.

9.6 Performance Evaluation

In this section, the effectiveness of ARA is validated while assuming a set of in-network heterogeneous nodes. Simulation results show the performance efficiency in terms of *average delay, price, idle time,* and *throughput* in comparison to key approaches in the literature. In addition, the quality of data delivery approach is assessed under varying rates per number of bytes transmitted, average energy consumption, and several counts of the network nodes.

9.6.1 Simulation Setup and Baseline Approaches

Using MATLAB R2016a and Simulink 8.7, we simulate randomly generated heterogeneous networks. The generated networks are random in terms of positions and densities of their nodes. To route data in these randomly generated networks, we apply our ARA scheme. The output of the ARA scheme is compared with the output of another four baseline approaches in the literature. These baseline approaches address the same problem tackled in this research; however, they use different routing strategies. The first approach forms a minimum spanning tree (MST) to find the most reliable route in a heterogeneous sensor network [29], and we call it MST approach; the second is for solving a Steiner tree (ST) problem with minimum number of Steiner points [30], and we call it ST; the third for adaptive data delivery, and we call it dynamic routing approach (DRA) [31], and the fourth one called LinGO [32]. In all these baselines, we assume a packet size equal to 512 bits, which is the typical size for IoT communication protocols. Every S node has an initial energy of 50 J and generates 150 pkts/round. A round is defined as the time span per which all S nodes have reported/requested a piece of data.

The MST opts to establish an MST through selected multihop paths. It first computes an MST for the given source and destination nodes and then forward messages over the minimum tree model; in which it finds the least count of hops to maintain the best-path cost. ST first combines nodes that can directly reach each other into one connected graph. The algorithm then identifies for every three connected graphs at node x that is at most r (m) away. Then, these three connected graphs are merged into one. These steps are repeated until no such x could be identified (i.e. no isolated nodes). DRA takes into consideration the nodes' coordinates to limit the updates sent out by the moving node to a local area. LinGO, which is a Link quality and Geographical beaconless OR protocol, introduces a different progress calculation approach compared with the aforementioned ones. It takes into account both the progress of a given forwarding node towards the destination with respect to the last hop, as well as the radio range. In this way, LinGO reduces the number of required hops. Both MST and ST routing strategies are used as a benchmark in this research due to their efficiency in finding the nearest next hop towards the destination while maintaining the minimum number of required nodes in the source–destination path. On the other hand, DRA is chosen due to its efficiency in adapting to any newly generated topology due to node mobility/heterogeneity. We remark that the original MST, ST, and DRA approaches are not hierarchical. Thus, we employed modified versions of them to make them suitable for our proposed hierarchical framework, where the modified versions take into consideration the in-network node heterogeneity and choose the next hop based on the type of surrounding node types. For example, a Zigbee-based IoT node will scan for another Zigbee-based node to be considered as a candidate neighboring node for packet relaying/forwarding. Since a larger network size implies longer paths, and thus, higher probabilities for heterogeneity. We examined the four data delivery schemes while the size of the network increases in terms of the count of IoT nodes. Knowing that larger node count in a data path raises the risk of node failure and, hence, dropped packets. Thus, choosing shorter peripheral paths is better for the overall quality/price gain.

The routing schemes, such as DRA, MST, ST, and ARA, are executed on 600 randomly generated wireless heterogeneous network topologies to get statistically stable results. The average results hold confidence intervals of no more than 2% of the average values at a 95% confidence level. We assume a predefined fixed time schedule for traffic generation at these networks. Data packets are delivered by applying these three approaches. Based on experimental measurements taken in a site of dense heterogeneous nodes [18], we set the communication model variables and other simulation parameters as shown in Table 9.3. We adopt the described signal propagation model in Section 9.3, where the utilized variables/parameters

Table 9.3   Parameters of the Simulated Networks

Parameter	Value
T	70%
n_c	110
Ψ	0.001 (ms)
D_{max}	500 (ms)
ω	200,000 (km/s)
P_l	4.8
δ^2	10
P_r	−104 (dB)
K_0	42.152
r	100 (m)

values, shown in Table 9.3, are set to be as follows: $P_l = 4.8$, $\delta = 10$, $Pr = -104$ (dB), and μ to be a random variable that follows a log-normal distribution function with mean 0 and variance δ^2.

Moreover, we assume heterogeneous transceiver communication ranges to validate our results in a typical IoT setup. For validation and verification purposes, we also used MATLAB with Simulink framework [33]. *Simulink* can support wireless channel temporal variations, node mobility, and node failures. The simulations last for 2 h and run with the log-normal shadowing path loss model [34]. In *Simulink*, we also adopted the same path loss and physical layer parameters shown in Table 9.3.

9.6.2 Performance Parameters and Metrics

To compare the performance of these three schemes, the following four performance metrics are used.

1. Average delay: It is measured in milliseconds and is defined as the average amount of time required to deliver a data unit to the destination.
2. Idle time: This metric reflects the ratio of idle time every node spends while just waiting to forward a message. It is measured in microseconds.
3. Throughput: This is set as a quality measure. It is the average percentage of transmitted data packets that succeed in reaching the destination, reflecting the effect of node heterogeneity and delay in IoT setups over the utilized data delivery approach.

4. Average price: This metric is used to observe the influence of utilized data delivery approach on the overall price to deliver a data unit from the source to destination on average.

Meanwhile, the three data delivery performances are assessed using the following three parameters:

1. The size of the network in terms of total node count. This reflects the application's complexity and the scalability of the exploited routing scheme.
2. Average energy consumption rate per data unit (π_i') as an indicator of the network power saving.
3. Cost (γ_i) to observe the influence of the charged price rate over the utilized data delivery approach.

9.6.3 Simulation Results

For a varying number of heterogeneous nodes (between 50 and 350) and deployment space (=1,200 km²), Figure 9.4 compares ARA approach with DRA, MST, and ST in terms of data delivery latency. It shows how ARA outperforms the other approaches under varying network size. Unlike the other approaches, exchanged messages using ARA do not show a rapid increment in the end-to-end delay while the network size is growing. This is because of the utilized utility function in Eq. (9.5) that has been considered by ARA approach. Although DRA is the most adaptive approach, its delay is increasing rapidly while the network size is increasing. However, MST and ST approaches are wore in terms of delay. It is also worth mentioning that when the network size is greater than or equal to 300, the ARA cannot improve anymore in terms of delay. However, it is obvious that ARA approach is achieving the lowest delay with respect to the varying network size.

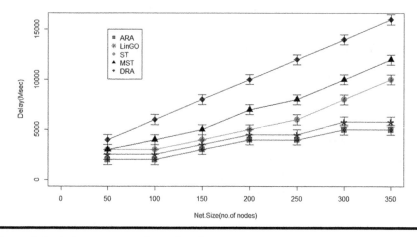

Figure 9.4 Latency vs. number of nodes.

Meanwhile, the overall network throughput levels achieved by ARA outperforms the levels achieved by other approaches due to considering the next hop status before forwarding the message to it, as depicted in Figure 9.5. In general, network throughput is increasing monotonically for all approaches while the network size is increasing. However, ST is the worst due to ignoring the current status of the node before message forwarding.

Furthermore, Figure 9.6 depicts the effects of the allowed average energy consumption level per node on the network throughput. It shows a monotonical increase in throughput for all approaches while varying the available energy budget. This comparison is performed while considering a fixed network size equal to 150 nodes, and average γ_i rate is equal to 0.002 \$/byte. Notably, more saving in

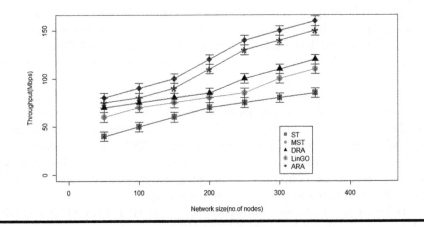

Figure 9.5   Throughput vs. the network size.

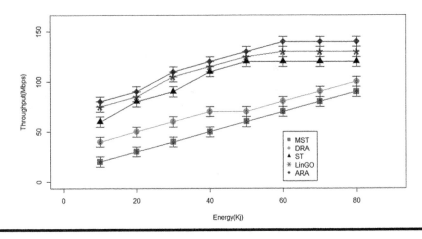

Figure 9.6   Throughput vs. the available energy.

terms of energy is achieved by applying the ARA and LinGO approaches. However, ARA approach achieves the highest throughput with respect to energy. When the energy budget is greater than or equal to 60 (kJ), the network throughput is saturated due to other design factors such as γ_i and network size and capacity. Also, it is worth to remark that LinGO adds redundant packets to increase the packet delivery probability while experiencing link error periods. This leads to significant increment in the overall throughput.

In Figure 9.7, the network end-to-end delay is decreasing linearly for all methods while γ_i is less than or equal to 50% of the initial rate. This can be returned to the main objective of all these methods in providing the best QoS while considering the cost factor. However, ARA again has the best delay performance with respect to all other approaches due to direct influence of the utility function in Eq. (9.5) in choosing the next hop towards destination. Also, it is worth noting that when γ_i is greater than or equal to 50%, all approaches cannot improve anymore in terms of end-to-end delay. This has a great impact on the network QoS.

In Figure 9.8, the average idle time is compared under varying total count of network nodes. All approaches are experiencing a monotonical increase in the average idle time while increasing the network size. This is expected due to the availability of several routing options/resources. In addition, we return the increase of idle time when the network size is increased due to the dense distribution of network nodes within a fixed deployment space ($=1200\,\text{km}^2$). Such a dense distribution provides idler resources as well. Nevertheless, ARA has the lowest average idle time in this comparison due to the ability of the proposed utility function in Eq. (9.5), where better resource management and utilization are guaranteed.

In Figure 9.9, ARA consistently outperforms the MST, ST, LinGO, and DRA while increasing the node counts. ST and DRA are the worst in terms of average price, which can be returned to their complexity in locating the next hop. Unlike

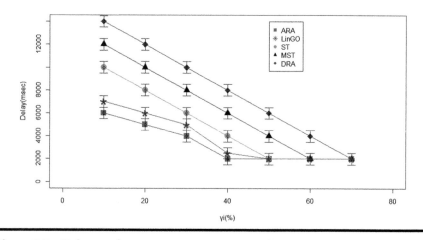

Figure 9.7 Delay vs. the average γ_i rate percentage.

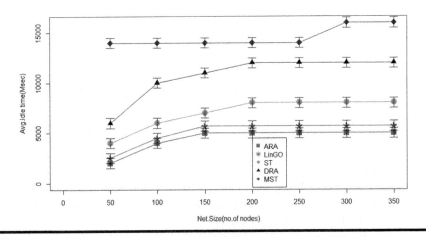

Figure 9.8   Avg. idle time vs. the network size.

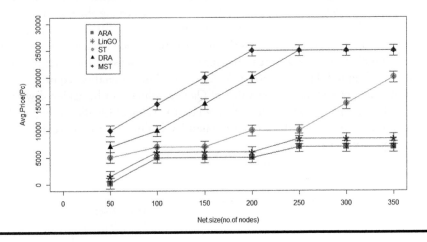

Figure 9.9   Avg. price vs. the network size.

ST and DRA, MST and LinGO approaches are very close to ARA. However, ARA is still outperforming them. The reason is that LinGO and MST adds redundant packets to increase the packet delivery probability while experiencing link error periods. This leads to significant increment in the overall price. In general, ARA is better because of the computed price factor based on the state of the current resources at every node n_i represented by Ψ_i.

It is also noted that the average price is increasing while the network size is increasing. Again, this has been accomplished due to considering longer routes while expanding the network size. We remark that as the node count increases the total price achieved by ARA becomes more identical and minimal with respect to other approaches. This can be returned to the excess in the available nodes, and thus, the effect of better choices becomes observable and prevents any increment in the price. In general, the function in Eq. (9.5) utilizes the aforementioned parameters after normalization, in a manner that maps the expected user experience to changes in individual utility parameters.

To show the impact of the aforementioned parameters on our utility function in Eq. (9.5), we present the plots in Figure9.10–9.12 for each of the utility parameters *Delay, Network Quality* (i.e. throughput), and Trust T, respectively. In Figure 9.10, *Delay* is plotted with a constant α that determines the rate of decrease of the exponential utility in Eq. (9.5). This particular function was chosen for *Delay* to reflect the rate of loss in the quality of e[1] as delay increases. By varying the value of α, it is possible to achieve different levels of delay tolerance as shown in Figure 9.10, where we chose $\alpha = 0.5$ for delay-tolerant data and $\alpha = 0.1$ for more delay-sensitive data. We note that, for a delay-sensitive data request, a very low delay has to be achieved to provide a high-delay utility component.

The quality parameter is plotted in Figure 9.11. We note that we adopt a Sigmoid function according to Eq. (9.5), where the tolerance to variation in data quality is expressed by fixing the value of ϵ (set here to 10) and varying the inflection point denoted by the value of β. Thus, if the requested data is quality sensitive (e.g. VoIP [voice over internet protocol] is sensitive to low transmission rate), the function will require a higher value before the utility increases (as depicted in the lower plot with $\beta = 0.8$ in Figure 9.11). In contrast, lower constraints on quality require a utility that increases rapidly at a lower value of *Quality*, which can be achieved with an early inflection point ($\beta = 0.5$ in Figure 9.11). The value of Quality in Figure 9.11 ranges from 0 to 1, where 1 indicates the best level of quality attainable depending on the quality metric.

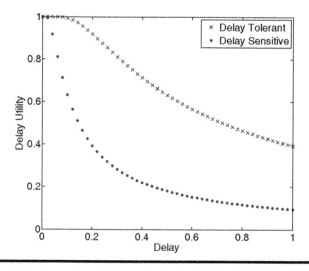

Figure 9.10 Delay function plot.

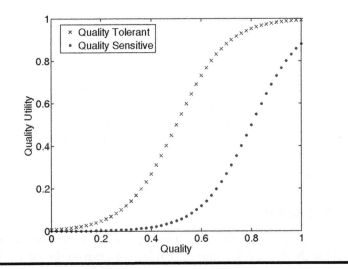

Figure 9.11 Quality function plot.

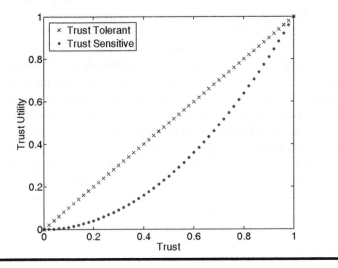

Figure 9.12 Trust function plot.

Lastly, Figure 9.12 shows the plot of Trust function, where $T \in [0,1]$. Note that an IN can give more emphasis to this parameter through the factor σ to particularly penalize sinks with bad service accounts. This is shown in the lower plot of Figure 9.12 where $\sigma = 2$, whereas the upper (better T) plot is a result of $\sigma = 1$.

We note here that we multiply the Trust component by the rest of the utility function in Eq. (9.5) to balance the effect of "deceiving" INs that may offer attractively low P_i to pass false quality promises. Moreover, we divide the utility function by $P_{intermediate}$ to protect the client from situations where two or more INs happen to achieve almost equal utility scores while charging prices that, although less than P_i, largely vary.

Evaluation results revealed that price-based routing scheme is better in most cases. Price-based routing scheme has a longer path length in contrast to shortest routing scheme as expected. But it has got better results in terms of cost, latency, idle time, and scalability. ARA avoids using busy, centric nodes, thus data in ARA deliver over more nodes. Data reach to destination over available nodes instead of waiting for queue in busy nodes.

9.7 Conclusion

This chapter proposed a price-based routing scheme for heterogeneous IoT networks called the ARA. ARA aims at establishing a cap on the internodal routing price to dynamically utilize the internet backbone if the source to destination distance surpasses a preset (case optimized) threshold. Promising simulation results are achieved by altering significant parameters including network size and the available energy budgets, and seeing how it affects certain metrics in the IoT paradigm such as latency, cost, and resource utilization.

These results show the efficiency of our framework when compared with three prominent ad hoc data delivery protocols. Our simulation results show that the ARA exhibits superior performance for different network sizes, lifetime, end-to-end delays, quality, and prices. It is strongly recommended with huge size networks as it is the most cost-effective one in the long run.

Future work would investigate using IoT nodes in smart city settings as mobile data collectors with semideterministic mobility trajectories. Also of practical interest is the application of localization methods among sensors operating on different technologies and studying the effect of such methods on the system's performance and the delivery rate of the corresponding ARA scheme.

References

1. F. Al-Turjman, A. Alfagih, and H. Hassanein, A novel cost-effective architecture and deployment strategy for integrated RFID and WSN systems, in *Proceedings of the IEEE International Conference on Computing, Networking and Communications (ICNC)*, Maui, Hawaii, 2012, pp. 835–839.
2. F. Al-Turjman, 5G-enabled devices and smart-spaces in social-IoT: An overview, Elsevier Future Generation Computer Systems, vol. 92, no. 1, 732–744, 2019.
3. L. Atzori, A. Iera, and G. Morabito, The internet of things: A survey computer networks, *Computer Networks*, vol. 54, 2787–2805, 2010. Elsevier.
4. G. Singh and F. Al-Turjman, Learning data delivery paths in QoI-aware information-centric sensor networks, *IEEE Internet of Things Journal*, vol. 3, no. 4, 572–580, 2016.

5. Q. Zhu, et al. IOT gateway: Bridging wireless sensor networks into internet of things, in *Proceedings of the IEEE/IFIP International Conference on Embedded and Ubiquitous Computing (EUC)*, Hong Kong, China, 2010, pp. 347–352.

6. F. Al-Turjman, Hybrid approach for mobile couriers election in smart-cities, *in Proceedings of the IEEE Local Computer Networks (LCN)*, Dubai, UAE, 2016, pp. 507–510.

7. E. Welbourne, et al. Building the internet of things using RFID: The RFID ecosystem experience, *Internet Computing, IEEE*, vol. 13, 48–55, 2009.

8. I. Fajjari, et al., New virtual network static embedding strategy within the cloud's private backbone network, *Elsevier Computer Networks*, vol. 62, 69–88, April 2014.

9. L. Boloni and D. Turgut, Should I send now or send later? A decision-theoretic approach to transmission scheduling in sensor networks with mobile sinks, *Wiley's Wireless Communications and Mobile Computing Journal (WCMC)*, vol. 8, no. 3, 385–403, 2008.

10. G. Wang, D. Turgut, L. Boloni, Y. Ji, and D. Marinescu, Improving routing performance through m-limited forwarding in power-constrained wireless networks, *Journal of Parallel and Distributed Computing (JPDC)*, vol. 68, no. 4, 501–514, 2008.

11. G. Solmaz, M. I. Akbas, and D. Turgut, A mobility model of theme park visitors, *IEEE Transactions on Mobile Computing (TMC)*, vol. 14, no. 12, 2406–2418, 2015.

12. D. Turgut and L. Boloni, Heuristic approaches for transmission scheduling in sensor networks with multiple mobile sinks, *The Computer Journal*, vol. 54, no. 3, 332–344, 2011.

13. F. Al-Turjman and H. Hassanein, Towards augmented connectivity with delay constraints in WSN federation, *International Journal of Ad Hoc and Ubiquitous Computing*, vol. 11, no. 2, 97–108, 2012.

14. M. Kranz, et al., Embedded interaction: Interacting with the internet of things, *Internet Computing, IEEE*, vol. 14, no. 12, 46–53, 2010.

15. A. Sarma and J. GirÃo, Identities in the future internet of things, *Wireless Personal Communications*, vol. 49, 353–363, 2009. Springer Netherlands.

16. M. Afergan, Using repeated games to design incentive-based routing systems, *in Proceedings of the IEEE International Conference on Computer Communication, (INFOCOM)*, Barcelona, Spain, 2006, pp. 1–13.

17. S. Zhong, et al., On designing incentive-compatible routing and forwarding protocols in wireless ad-hoc networks, *Wireless Networks*, vol. 13, no. 9, 799–816, 2007.

18. X. Wang and H. Schulzrinne, Pricing network resources for adaptive applications, *IEEE/ACM Transactions on Networking*, vol. 14, no. 3, 506–519, June 2006.

19. G. V. Ozianyi, N. Ventura, and E. Golovins, A novel pricing approach to support QoS in 3G networks, *Computer Networks*, vol. 52, no. 7, 1433–1450, May 2008.

20. F. Al-Turjman, Cognition in information-centric sensor networks for IoT applications: An overview, *Annals of Telecommunications*, 2016, pp. 1–16. doi:10.1007/s12243-016-0533-8. Springer.

21. W. Saad, Z. Han, M. Debbah, A. Hjorungnes, and T. Basar, Coalitional game theory for communication networks, *Signal Processing Magazine, IEEE*, vol. 26, 77–97, 2009.

22. G. Singh and F. Al-Turjman, A data delivery framework for cognitive information-centric sensor networks in smart outdoor monitoring, *Computer Communications*, vol. 74, no. 1, 38–51, 2016. Elsevier.

23. H. Sundmaeker, P. Guillemin, P. Friess, and S. Woelfflé, Vision and challenges for realising the Internet of Things, *in CERP-IoT, European Commission*, Luxembourg, 2010.

24. W. Heinzelman, A. Chandrakasan, and H. Balakrishnan, Energy-efficient communication protocols for wireless microsensor networks, *in Proceedings of Hawaiian International Conference on Systems Science*, Big Island, Hawai, 2000, pp. 2–10.
25. A. DaSilva, Pricing for QoS-enabled networks: A survey, *IEEE Communications Surveys and Tutorials*, vol. 3, no. 2, pp. 2–8, 2000.
26. F. Al-Turjman, and S. Alturjman, 5G/IoT-enabled UAVs for multimedia delivery in industry-oriented applications, Springer's Multimedia Tools and Applications Journal, 2018. doi:10.1007/s11042-018-6288-7.
27. F. Al-Turjman, Cognitive routing protocol for disaster-inspired internet of things, *Future Generation Computer Systems*, vol. 92, 1103–1115, 2019.
28. M. Z. Hasan, et al. A survey on multipath routing protocols for QoS assurances in real-time multimedia wireless sensor networks, *IEEE Communications Surveys and Tutorials*, vol. 19, 1424–1456, 2017. doi:10.1109/COMST.2017.2661201.
29. F. Al-Turjman, H. Hassanein, and M. Ibnkahla, Towards prolonged lifetime for deployed WSNs in outdoor environment monitoring, *Ad Hoc Networks Journal*, vol. 24, no. A, 172–185, January 2015. Elsevier.
30. F. Al-Turjman, A novel approach for drones positioning in mission critical applications, Wiley Transactions on Emerging Telecommunications Technologies, 2019. doi:10.1002/ett.3603.
31. F. Al-Turjman, Fog-based caching in software-defined information-centric networks, Elsevier Computers & Electrical Engineering Journal, vol. 69, no. 1, 54–67, 2018.
32. D. Rosário, Z. Zhao, A. Santos, T. Braun, E. Cerqueira, A beaconless Opportunistic Routing based on a cross-layer approach for efficient video dissemination in mobile multimedia IoT applications, *Computer Communications*, vol. 45, 21–31, 2014. Elsevier.
33. https://mathworks.com/products/simulink/. SIMUTOOLS2010.8727. doi: 10.4108/ICST.SIMUTOOLS2010.8727.
34. M. Z. Hasan, et al., Optimized multi-constrained quality-of-service multipath routing approach for multimedia sensor networks, *IEEE Sensors Journal*, vol. 17, no. 7, 2298–2309, 2017.

Chapter 10

Security in UAV/Drone Communications

Fadi Al-Turjman and Jehad M. Hamamreh
Antalya International (Bilim) University

10.1 Introduction

The rapid advancement in internet of things (IoT) technology enabled connectivity to a large number of smart devices, where they can be accessed anytime, anywhere, and by everyone [1,2]. Meanwhile, drone technology, known as unmanned aerial vehicle (UAV), witnessed a vast attention in the recent years as well due to the tremendous advantages they can offer and their deployment flexibility. To this extent, both technologies form a promising paradigm that offers a wide range of applications in smart spaces known as internet of drones (IoD).

UAV (drone)-based communication is becoming one of the key promising applications of UAV systems, which can also be used for other inherited applications such as surveillance, tracking, transportation, environmental monitoring, industrial automation, agriculture, public safety, delivery, filmography, disaster relief (search and rescue), air exploration, target localization, fighting, etc. These enabled applications of UAVs and many others are attributed to their key features and characteristics, including aerial mobility with adaptive altitude, changeable location and direction, easy deployment, expandability, flexibility, and adaptive usage [1–4].

Particularly, UAV-based communication is becoming not only an achievable reality with many new benefits but also a key potential solution to a number of communication and networking challenges that may result due to natural disaster scenarios. In fact, the 3rd Generation Partnership Project (3GPP) standardization community has identified and specified several possible deployment scenarios for UAV-based communication in the domain of 5G systems, as detailed in the standardization documents named TS 22.261, TR 22.862, and TR 36.777 [4,5]. Specifically, UAV can be utilized as (1) an aerial base station (BS) providing connection links to multiple terrestrial or aerial users, (2) an aerial IoT component, which is basically a user equipment (UE) flying in the air, and (3) an aerial relay that can be used to handover data traffic from one point to another. In addition, it can be used as a flying jammer to help enhance the security of certain scenarios.

Regardless of the deployment scenarios, there are several key, essential and vital, requirements that have to be met to ensure the successful usage of UAV communication technology. Among the many design requirements of UAVs, especially for ultrareliability and low-latency communication -based 5G services, we mention low complexity, high reliability, high energy efficiency, low latency, and robust security [6]. Among these design requirements, communication security comes as one of the most critical and important key priority to fulfill, to guarantee the successful deployment of UAV systems. To meet this design goal, novel security algorithms are needed.

Generally speaking, UAVs should by default be able to satisfy the following conditions and requirements for achieving an acceptable level of security:

■ Confidentiality: The confidentiality requirement for UAV communication, or what is known in the literature as the eavesdropping problem, refers to the situation where a legitimate transmitter (Alice) tries to communicate secretly with another legitimate/intended user (Bob) under the presence of a third unauthorized/unintended user called eavesdropper (Eve), who tries to intercept and overhear the communication content between the legitimate parties (Alice and Bob). Thus, the primary objective of confidentiality-guaranteeing algorithms (eavesdropping-resilient methods) is to limit data access to intended users only, while preventing the disclosure and leakage of information to unintended, malicious eavesdroppers.

■ Authentication: Reactions given by UAVs to events should be based on legitimate messages. Therefore, proper lightweight protocols of authentication should be employed by the sender and/or receiver, either in public or private networks.

■ Plausibility: The legitimacy of transmitted messages also includes the evaluation of their consistency with similar ones, as the legitimacy of the broadcaster can be assured while the contents of the message contain erroneous data. The way that the plausibility is confirmed will firmly depend on the type of data transferred.

■ Availability: Even when we assume the existence of a robust communication channel, some attacks and malfunctions can weaken and bring down the network by finding flaws in the system. Therefore, it is paramount that availability of UAV services should also be supported by alternative means.

This can be achieved via a UAV to UAV or a UAV to infrastructure solution with a backup protocol in place.

- Nonrepudiation: UAVs causing illegal actions need to be reliably identified while the sender should not be able to pick and choose which message to broadcast or deny for a certain message.
- Time criticality: Considering the mobility factor in a typical UAV network, stringent constraints should be expected when dealing with time-sensitive data, as it might not leave any room for mistakes and could have disastrous results if not met properly.
- Privacy: The privacy of UAVs and their messages against unauthorized observers have to be guaranteed. This is a chief concern as the development of UAVs is following a customer-based demand that cannot be realized unless the customer privacy is guaranteed.
- Trust: The primary element in any secure UAV system is trust and privacy. This is particularly true and critical in UAVs due to the high liability expected in their safety and security applications and, consequently, the members running them. With a significant number of independent nodes in the UAV network and the presence of human factor, it is without a doubt highly probable that misbehavior can occur. In a connected IoT-based space, users are increasingly concerned about their privacy, and UAVs are by no means an exception. This is especially problematic as the lack of privacy and the potential tracking functionality inherent in UAVs can lead to severe privacy violence. Accordingly, UAVs and service providers must be mutually controlled by a considerable presence of governmental authorities.

Among the aforementioned security requirements for UAV communications, preserving data confidentiality, i.e., providing security against eavesdropping by allowing data access to only authorized/legitimate users, while forbidding unauthorized/unintended users (called eavesdroppers) from intercepting the information, comes at the highest priority. This is because of the fact that guaranteeing data confidentiality provides a first line of defense against not only eavesdropping but also against many other attacks such as denial of service (DoS), data modification, man-in-the-middle (MITM), session hijacking, spoofing (impersonation), and sniffing.

In the context of drone communications, providing confidentiality for legitimate receivers against unintended ones (eavesdroppers) appears to be a challenging goal to achieve due to the unique transmission characteristics and nature of UAV systems, including broadcasting, dominant line of site, and poor scattering. Besides, it is believed that due to having strict requirements on the power, weight, and processing capabilities of UAVs, complexity-based cryptographic algorithms [7–9] cannot be supported by the BS carried by the UAV, which is unlike a terrestrial BS that has enough processing capabilities to support sophisticated encryption schemes. As a result, light cryptography is considered as a potential approach to provide security while reducing complexity, which results in power saving that can be reused for operating UAVs for longer required period of time. However, this approach comes at the expense of reducing the security level as it becomes easier for eavesdropper to perform hacking due to the fact the encryption algorithms are light and not complex. Consequently, this approach, although saves power and reduces complexity, can make the UAV susceptible to security threats and vulnerabilities.

This particular problem motives the use of physical layer security (PLS) approaches for securing UAV-based systems against eavesdropping due to their complexity-independent secrecy. This is so because no matter what computational power and processing complexity the eavesdropper may have, there is no feasible way to decrypt the security algorithms [10–17].

This chapter is dedicated to highlight and overview the latest advances and state-of-the-art research attempts performed towards applying PLS to UAV communication systems. This is performed by classifying the existing research studies according to the deployment scenarios and use cases of UAVs (i.e. UAV-relay, UAV-BS, UAV-UE, UAV-jammer). Also, we discuss and explain some of the additional common attacks in UAV systems. Then, we propose recommendations and future research directions.

10.2 PLS for UAV Systems

The classical solutions that are being used to deliver secure communication in UAV-based systems are based on cryptography approaches similar to other wireless technologies. However, conventional complexity-based encryption algorithms are deemed unsuitable for future technologies including UAV communication systems due to the following practical reasons: First, future networks are composed of heterogeneous and decentralized wireless access technologies, where key distribution, management, and maintenance processes are deemed very difficult and challenging in such scenarios. Second, future networks need to support new wireless technologies like IoD to enable many diverse applications. The transceiver devices in these wireless technologies are naturally (1) power-limited due to depending on battery sources, (2) processing-restricted due to having low computational capabilities in terms of random access memory (RAM), central processing unit (CPU), and storage unit, and (3) delay-sensitive due to using control-based applications. All these facts together make cryptography-based techniques infeasible and ineffective for such type of technology. Third, future networks are expected to support diverse services, applications, and scenarios with different levels of security requirements. However, encryption-based algorithms lack the ability to deliver different levels of security.

To cope up with the aforementioned issues, the PLS concept has emerged as a promising solution that is capable of addressing some of the hurdles associated with cryptography. PLS exploits the dynamic nature of wireless channel along with its features, including randomness, location dependency, fading, dispersion, spreading, interference, noise, etc., to prohibit the eavesdropper from decoding the received data, while guaranteeing that the legitimate user can decode it successfully.

These aforementioned facts and reasons motivate the use of PLS as a promising solution to address security concerns in UAVs systems. In this section, we discuss and overview the latest advances in this area.

10.2.1 UAV as a Mobile Relay (UAV Relay)

The first work in the literature that studies using drones to improve the PLS of wireless communication systems is conducted by Wang et al. [18]. In this work, the authors maximize the secrecy rate of a system composed of four nodes: a transmitter source (Alice), a UAV (mobile relay), a ground receiver destination (Bob), and an eavesdropper (Eve) located near Bob, as shown in Figure 10.1. The authors show that the resulting secrecy problem formulation of such scenario is nonconvex and hard to solve. Therefore, an iterative algorithm is developed by exploiting the difference of concave program to solve the optimization problem.

The obtained results show that the use of mobile relaying enabled by UAV can significantly give better performance in terms of secrecy than the use of static relaying [18].

Similar to the previous study [18], the authors in [19] investigate the transmission secrecy of a system composed of a four-node setup (source, destination, mobile relay, and eavesdropper). Specifically, the secrecy rate of the system is maximized by performing joint optimization for both: trajectory of the relaying UAV and the transmitted power of source and relay. It is shown that the resulting secrecy rate maximization problem is difficult to solve due to the direct influence between power allocation and trajectory optimization. Therefore, an alternating optimization strategy is proposed, by which the trajectory design and power allocation are handled in an alternating way. Sequential convex programming is also used to overcome the nonconvexity problem of trajectory optimization, thus enabling the derivation of an iterative, convergent algorithm. The obtained results verify the

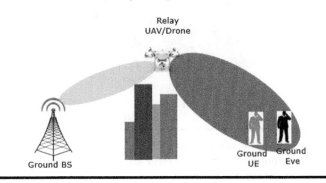

Figure 10.1 A communication scenario composed of a ground transmitter BS, relay UAV, ground legitimate receiver UE, and ground eavesdropper (Eve).

effectiveness of the proposed joint power and trajectory optimization in enhancing the secrecy performance.

In [20], an effective security scheme is introduced to guarantee the security of UAV-relayed wireless networks against eavesdropping with caching via jointly optimizing the UAV trajectory and time scheduling.

The authors of [21] study the secrecy outage performance resulting from using opportunistic relaying for a low-altitude UAV swarm[1] in the presence of multiple UAV eavesdroppers. Particularly, multiple UAV transmitters, which are served by a ground BS and multiple UAV relays, are optimally selected to help enhance the secrecy of transmitted confidential messages to a far ground user under the presence of multiple flying eavesdroppers.

10.2.2 UAV as a Mobile Transmitter BS (UAV-BS)

Unlike the aforementioned work that uses UAV as a mobile relay to enhance secrecy, Zhang et al. [22] consider the PLS of a system composed of a UAV node (Alice) that acts as a mobile transmitter BS (UAV-BS) and sends secret information to a legitimate receiver (Bob) located on the ground in the presence of an eavesdropper, who is also situated on the ground as depicted in Figure 10.2. The authors maximize the secrecy rate of the aforementioned system setup by using joint optimization of the transmit power and trajectory of the mobile UAV over a finite horizon. The formulated nonconvex optimization problem of the aforementioned system setup

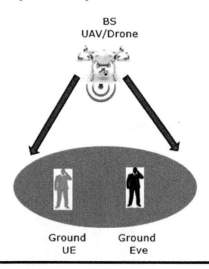

Figure 10.2   A communication scenario composed of an aerial, flying transmitter BS (UAV-BS), ground legitimate receiver UE, and ground eavesdropper (Eve).

---

[1] UAV swarm is a communication engineering term, which is similar to the flying ad hoc network term used in networking literature.

is solved by an iterative algorithm that is based on successive convex optimization and block coordinate descent methods. The presented results in [22] demonstrate the capability of an algorithm to significantly enhance the secrecy rate of the UAV system, when compared with other schemes that neither consider transmit power control nor trajectory optimization.

In [23], the authors target to enhance the PLS performance of a system that consists of a mobile UAV-BS (Alice) communicating with a ground receiver node (Bob) under the presence of K number of potential eavesdroppers (Eves), who are located on the ground as well, and their location information is imperfect at the UAV-BS. To achieve this, the authors formulate an optimization problem to maximize the average worst-case secrecy rate of the system through designing the robust trajectory and transmit power of the UAV over a given flight duration. The resulting optimization problem is shown to be hard to solve optimally due its nonconvexity from one hand and the imperfect location information of the eavesdroppers on the other hand.

Therefore, an iterative suboptimal algorithm is proposed to tackle this problem effectively by using the S-procedure algorithm, block coordinate descent method, and successive convex optimization method. Numerical results show a noticeable, significant improvement in the average worst-case secrecy rate using the proposed design in comparison with other designs that do not consider joint optimization.

The downlink (UAV-to-ground) and uplink (ground-to-UAV) communications with a ground node, subject to an eavesdropper located on the ground, is considered in [24]. In this study, the high mobility of the UAV alongside its trajectory design is exploited to create a good-quality channel for the legitimate link, and a degraded (low-quality) channel for the eavesdropping link. New problems are formulated to maximize the average secrecy rates of the downlink and uplink transmissions via jointly optimizing the transmit power of the legitimate transmitter and the trajectory of the UAV. Iterative algorithms are proposed to effectively solve the formulated problems as they are found to be nonconvex. This is attained using the successive convex optimization and block coordinate descent methods. The acquired results exhibit performance enhancement in the secrecy rates by the proposed algorithms, in comparison to other reference designs that neither use trajectory optimization nor power control.

The integration between UAV and mm-wave systems has recently been studied in the literature. Particularly, the PLS aspect of this integration has been investigated and analyzed in a recent work performed by Zhu et al. [25]. In this work, the authors consider a downlink mm-wave network composed of multiple UAVs that serve and work as aerial, flying BSs to provide wireless converge and connectivity to multiple legitimate receivers on the ground, which are surrounded by multiple eavesdroppers.

10.2.3 *UAV as Mobile Jammer (UAV-Jammer)*

Besides using UAV as a mobile BS [22] or as a mobile relay [18], Zhang et al. [26] propose the use of UAV as a jammer to improve communication secrecy. Particularly, they consider a scenario in which a source BS on the ground (Alice) communicates with a legitimate receiver (Bob), who is also located on the ground, whereas an eavesdropper (Eve) tries to intercept the ongoing legitimate transmission link [26] as shown in Figure 10.3. On the other hand, a UAV-based jammer is considered to be deployed in the system setup to improve the secrecy performance by emitting intelligent artificial noise.

The authors of [27] introduce an effective secrecy scheme that uses a UAV as a mobile jammer to enhance the secrecy rate of a ground wiretap channel. In particular, a UAV is employed to protect transmission against eavesdropping by transmitting intelligent jamming signals that can affect an eavesdropper more than the legitimate receiver, as the UAV-enabled jammer can move away from the legitimate receiver so that it can get closer to the eavesdropper (if its location is known). The approach here is to jointly optimize the jamming power and UAV trajectory to maximize the average secrecy rate. A closed-form lower bound on the achievable secrecy rate is derived to make the problem analyzable and more tractable. By using this bound, the transmit power and UAV trajectory are optimized alternately by employing an iterative algorithm that uses successive convex optimization and block coordinate descent techniques. Numerical results demonstrate significant improvements in the secrecy rate of the considered wiretap system by the adopted joint design when compared with other nonoptimized schemes in the literature.

The utilization of UAV nodes for the benefit of cognitive radio communication seems to be an effective solution for certain challenges. Most importantly, the PLS aspect of this utilization has been introduced [28]. In this work, PLS is considered for cognitive radio networks using UAV-enabled jamming noise. Specifically, a secondary transmitter sends confidential messages to a secondary receiver in the presence of an external eavesdropper (Eve), and the UAV acts as a friendly jammer that degrades the decoding capability of Eve. To maximize the secrecy rate of such a scenario while guaranteeing a certain signal-to-interference threshold at the

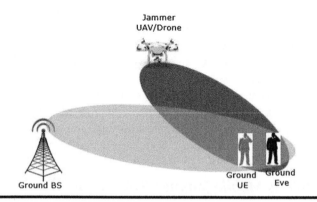

Figure 10.3 A communication scenario composed of a ground transmitter BS, UAV-jammer, ground legitimate receiver UE, and ground eavesdropper (Eve).

primary receiver, resource allocation has to jointly optimize the transmit power and trajectory of UAV. The resulting design problem is found to be nonconvex; therefore, in an attempt to solve the problem, it is proposed to convert the problem into a tractable form, and then use an effective, feasible, and low-complexity algorithm based on successive convex approximation. The obtained results verify the superiority of the proposed solution, compared with other available ones.

10.2.4 UAV as a Flying UE (UAV-UE)

The general goal in this category is to study the PLS of a system consisting of a ground BS transmitter (Alice) that acts as a control center/BS, and a UAV (Bob) that represents a flying UE (UAV-UE) in the presence of a flying eavesdropper (UAV-Eve), as shown in Figure 10.4. In [29], directional modulation (DM) is utilized by Alice to improve the secrecy rate performance of a system similar to the aforementioned one.

To further enhance the secrecy level, an alternating iterative structure between power allocation and beamforming is proposed to be employed by the system. Simulation results demonstrate that the proposed scheme can achieve substantial secrecy rate gains. Particularly, in the case of small-scale antenna array, the gain of the secrecy rate performance achieved by the proposed scheme is very significant.

The uplink (ground-to-UAV) and downlink (UAV-to-ground) communications with a ground node, subject to an eavesdropper located on the ground, is considered in [24]. In this study, the high mobility of the UAV alongside its trajectory design is exploited to create a good-quality channel for the legitimate link, and a degraded (low-quality) channel for the eavesdropping link. New problems are formulated to maximize the average secrecy rates of the downlink and uplink

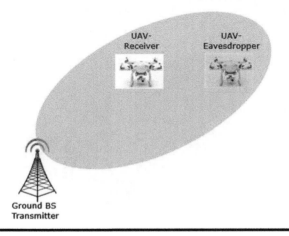

Figure 10.4   A communication scenario composed of a ground transmitter BS, UAV receiver UE, and UAV eavesdropper (Eve).

transmissions via jointly optimizing the transmit power of the legitimate transmitter and the trajectory of the UAV. Iterative algorithms are proposed to effectively solve the formulated problems as they are found to be nonconvex. This is attained using the successive convex optimization and block coordinate descent methods. The acquired results exhibit performance enhancement in the secrecy rates by the proposed algorithms, in comparison to other reference designs that neither use trajectory optimization nor power control.

10.2.5 One UAV as a Cooperative Jammer and Another as a Transmitter

To improve the secrecy performance of a UAV-based communication system supporting ground users, it is possible to utilize one UAV as a mobile cooperative jammer and another UAV as a source BS transmitter as shown in Figure 10.5.

In this direction, the authors of [30] present a UAV-aided mobile jamming strategy to further improve the achievable average secrecy rate for UAV-ground communications. Specifically, an extra cooperative UAV is used as a mobile jammer to broadcast jamming signals that can help keep the source UAV transmitter closer to the ground receiver, thus producing a good-quality legitimate link that results in enhancing the secrecy performance of the system. The design objective is achieved by maximizing the achievable secrecy rate through jointly optimizing the transmit power and trajectories of both jammer UAV and source UAV. An iterative algorithm based on a block coordinate descent method is used to solve the considered nonconvex optimization problem.

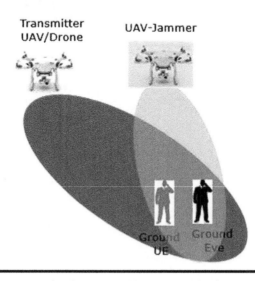

Figure 10.5 A communication scenario composed of a UAV transmitter BS, UAV-jammer, ground receiver UE, and ground UAV eavesdropper (Eve).

In [31], the authors study a scenario in which a mobile UAV tries to broadcast secret messages to multiple ground users. To enhance the secrecy performance of the system, a cooperative UAV that acts as a jammer is considered. In this setup, the lowest secrecy rate of the ground users is maximized by jointly optimizing the transmit power and the trajectory of the UAVs as well as the scheduling of the user. Block successive minimization techniques are adopted to efficiently solve this nonconvex problems.

In another work [32], the authors introduce an effective cooperative jamming approach to secure the UAV communication by utilizing jamming from other nearby UAVs to provide confidentiality and defend against external eavesdropping. Particularity, the authors consider a scenario composed of two UAVs, where one UAV acts as a source transmitter trying to send confidential information to a ground node (GN) and another UAV acts as a jammer that transmits artificial noise to confuse the ground eavesdropper.

The two UAVs can flexibly adopt and modify their trajectories (locations over time) to enable secure communication that is leveraging and exploiting not only the fully controllable mobility of UAVs but also the ability to use them as cooperative jammers. It is assumed that the location of the ground nodes is perfectly known by the two UAVs, whereas the eavesdropper's location is partially known ahead of time.

The design goal in this study is to maximize the average secrecy rate from the UAV transmitter to the ground node within a certain period of time. This is attained by optimizing the UAV trajectories, jointly with their communicating/jamming power allocations. Again, similar to other optimization secrecy problems in UAV-based scenarios, the formulated problem is found to be nonconvex, and thus a numerical solution is proposed by utilizing alternating optimization and successive convex approximation techniques.

10.3 Additional Common Attacks in UAV Systems

A general classification for both attackers and attacks identified in UAV-enabled networks is provided in this section.

10.3.1 Attacker Classification

Understanding the nature of the attacker is important for classifying the types of attacks that UAVs might be subject to. Attackers in UAV networks can be classified into the following:

■ Insider/Outsider: An authenticated UAV member of the network who can broadcast and receive messages from other members is an insider. It has access to a certified public key and can more easily navigate the network protocols for mounting attacks. Meanwhile, the outsider is considered as a foreign object by the UAV network members and as an intruder. Hence, it is severely limited in interactions by the applied security protocols.

■ Malicious/Rational: A malicious attacker is not in search of any kind of personal benefits from the attacks he perpetrates and aims to harm. The main issue is that this type of attackers can employ any means disregarding corresponding costs and consequences making the attacker unpredictable and potentially dangerous. On the contrary, an attacker that seeks personal profit is rational and will not overextend the resources for any intangible gain. This makes this type of attackers more predictable regarding the attack means and the attack targets.

■ Active/Passive: Active attackers generate intruding and misleading signals that can help in discovering the identity and/or unauthorized data access. Unlike, the active attacker, a passive attacker inserts himself with eavesdropping in the wireless channel.

■ Local/Extended: An attacker can be limited in scope, even if he/she controls several nodes, which makes him/her local, limiting his/her impact at large. The extended attacker can control several entities that are scattered across the UAV network, allowing him/her to be active on a larger scope. The differentiation between local and extended is not easy to make and depends heavily on the size and coverage of the UAVs in question. This distinction can be especially important in tracking privacy-violating activities and potential suspects.

10.3.2 Attack-Type Classification

Security attacks in UAVs can be classified into the following:

■ Masquerade: The UAV is actively masking its own identity to appear like another UAV by using false identities, such as public keys. This technique is usually employed in conjunction with other types of attacks.

■ False Information: A UAV can transmit erroneous information and data in the network, which might affect the behavior of other UAVs. It can be both intentional and unintentional.

■ Location Tracking: The observer can monitor the trajectories of selected UAV members and can use this information for a range of purposes, both malicious and mundane. It can also potentially leverage on the roadside units (RSUs) or UAVs that are around its main target.

■ DoS: An attacker can break down a UAV network, jam signals, or may even cause a collision by using malicious nodes to forge a significant number of bogus identities, such as IP (internet protocol) addresses, with the final objective of disrupting the proper functioning of data and information transfer between two moving UAVs. An example would be jamming the communication channel.

■ global positioning system (GPS) Spoofing: GPS satellites, or their equivalent, maintain a locational table with the spatial locations and identifiers of UAVs in the network. Attackers may produce misleading and false readings in the positioning system with the purpose to deceive UAVs, leading them to assume that they are actually in a different spot. It is relatively easy to dupe any amount of UAV members with some restrictions. It is also quite possible to use a GPS satellite simulator to generate and broadcast signals that are stronger than those generated by the actual satellite system, leading the UAV receivers to prefer it over the actual satellite.

■ Physical/Electronic Outage: When a GPS signal disappears in an outage due to obstacles, it can be exploited using the temporary loss of connection to the system to inject falsified data and positioning information. Once the UAV has a line of sight (LoS) with the GPS satellite, it assumes that this is the actual position and will tread as such before it receives an updated position from the satellite. A nonphysical outage could be created using proper channel jamming that broadens the potential applications of such attack.

■ Wormhole: Traditionally, this is accomplished by tunneling packets between two remote members of a UAV network. The perpetrator should control at least two nodes separate from each other and with a very high-speed connection between them to tunnel packets from one location to broadcast them in another. This could be accomplished with preestablished RSUs or by using fast mobile technologies such as 4G or 5G. Wormholes allow the attacker to spread misleading but properly signed messages at the destination area. A way to protect the UAV network from wormhole attacks is the Timed Efficient Stream Loss Tolerant Authentication (TESLA) with instant key disclosure. Another potential solution discussed in the literature is the AODV (ad hoc on-demand distance vector) routing protocol, where hop-by-hop efficient authentication protocol is a useful approach that allows us to notice wormhole attacks.

■ Black Hole: Data packets get lost while crossing through a black hole, which in effect is a member that has some nodes or no node that refuse to broadcast or forward data packets to the next hop. Preventing black hole attacks is generally achieved by making use of redundant paths kept between the sender and the receiver of the message. Another potential way to defend against black holes is the use of an information-carrying sequence number in the message header. The receiver can then potentially figure out the absence of a packet in the case of any discrepancy or loss, identifying the situation as a suspicious case.

■ Malware and Spam: Attacks like spam and viruses can lead to severe disruptions in UAV operations. They are typically the work of malicious insiders rather than outsiders who have access to UAVs and RSUs when they are performing software updates. They can potentially result in an increase in transmission latency, which can be lessened using a centralized management.

Proper maintenance of infrastructure and a centralized administration should be employed to prevent such attacks.

■ MITM: Malicious attackers can eavesdrop on the communication between UAVs and inject false information or distort messages between them. Solutions so far are relying mainly on strong cryptography, secure authentication, and data integrity verifications. However, that is not enough to prevent such attacks.

■ Illusion Attack: When an attacker broadcasts warning messages that do not correspond to the current situation, it produces an illusion to the UAV members in their neighborhood. The propagation of the phantasm mainly depends on the UAV's responses, which can lead to data traffic jams and a general degradation in the system behavior.

■ Impersonation Attack: During a UAV-to-UAV communication, a member can broadcast the security messages as if it was the origin to other UAVs that can potentially have an impact on the behavior of the network control and the other UAV members of the network. For example, a malicious UAV transmits a message on behalf of another member to create chaos or any other security attack while masquerading. To protect the identity of each UAV, a pseudoidentification code and a shared secret key between UAVs in the network can be used. This type of attacks mainly disturbs the identity keys, and it is paramount for identifying the origin of broadcasted messages.

10.4 Open Research Issues

According to the analysis on literature, we present the following open research gaps that future efforts may consider:

■ Extension of the PLS studies to multilink and multinode scenarios under the effect of different channel conditions and network topologies.
■ Development of resilient PLS schemes to protect the transmission not only from passive attacks such as eavesdropping but also from active ones such as spoofing and jamming.
■ The integration of drone/UAV technology with other emerging radio access communication technologies such as mm-wave and visible light communications is worth investigation and analysis to understand the capability of UAVs in enhancing the security of such high-frequency technologies.
■ Facilitation of random and controlled deployment methodologies enhances network security and user privacy.
■ Development of integrated security optimization schemes that consider trade-offs between data fidelity and confidentiality.
■ Secure design of the drone-based hardware and software used in critical applications.
■ Assessment of medium blockage effects on drone security.

- Simultaneous integration of different security paradigms to achieve more reliable performance.
- Assessment of surrounding conditions that affect drone security against external attacks.
- Design of data-intensive and time-sensitive secured drones, especially for emergency cases where drone efficiency is crucial.
- Optimization of dynamic drone trajectory in secured spaces for more trusted solutions.
- Facilitation of automatic collision avoidance in drones against any unexpected attack.
- Enforcement of drone-embedded security techniques to avoid reporting false locations and unauthorized data.

10.5 Conclusion

Due to the rapid technological advances and unprecedented growth in the number and type of flying vehicles, many applications using UAVs have emerged. Regardless of these UAV-enabled applications, which have different set of requirements and performance targets to meet in terms of reliability, latency, coverage, spectral and energy efficiency, etc., communication and networking security come as the most critical and important objective to guarantee safe operation and utilization of UAV-based systems.

However, due to the unique transmission characteristics and nature of UAV systems, including broadcasting, dominant line of site, and poor scattering, security appears as a challenging goal to meet in such scenario. Besides, the special features of UAVs represented by having limitation on power (run by battery) and processing (light RAM and CPU capabilities) make applying complex cryptography approaches very challenging and ineffective for such systems.

This has motived the use of PLS-based approaches for securing UAV-based systems due to their complexity-independent secrecy, as no matter what complexity the eavesdropper may have, there is no way to decrypt the security algorithms. This chapter has highlighted and overviewed (in a structured and unified manner) the latest advances and state-of-the-art researches in the field of applying PLS to UAV systems under different use cases and scenarios such as utilizing the UAV as a BS, relay, UE, and/or jammer. In addition, different types of attacks in UAV systems alongside future research directions have been identified and discussed.

References

1. L. Gupta, R. Jain, and G. Vaszkun, Survey of important issues in UAV communication networks, *IEEE Communications Surveys and Tutorials*, vol. 18, no. 2, 1123–1152, Secondquarter 2016.

2. S. Hayat, E. Yanmaz, and R. Muzaffar, Survey on unmanned aerial vehicle networks for civil applications: A communications viewpoint, *IEEE Communications Surveys and Tutorials*, vol. 18, no. 4, 2624–2661, Fourthquarter 2016.
3. R. Shakeri, M. A. Al-Garadi, A. Badawy, A. Mohamed, T. Khattab, A. K. Al-Ali, K. A. Harras, and M. Guizani, Design challenges of multi-UAV systems in cyber-physical applications: A comprehensive survey, and future directions, *CoRR*, vol. abs/1810.09729, 2018. [Online]. Available: http://arxiv.org/abs/1810.09729
4. A. Fotouhi, H. Qiang, M. Ding, M. Hassan, L. G. Giordano, A. Garc'ia-Rodr'iguez, and J. Yuan, Survey on UAV cellular communications: Practical aspects, standardization advancements, regulation, and security challenges, *CoRR*, vol. abs/1809.01752, 2018. [Online]. Available: http://arxiv.org/abs/1809.01752
5. F. Al-Turjman, and S. Alturjman, 5G/IoT-enabled UAVs for multimedia delivery in industry-oriented applications, *Springer's Multimedia Tools and Applications Journal*, 2018. doi:10.1007/s11042-018-6288-7.
6. J. M. Hamamreh, Z. E. Ankarali, and H. Arslan, CP-Less OFDM with alignment signals for enhancing spectral efficiency, reducing latency, and improving PHY security of 5G services, *IEEE Access*, vol. 6, 63649–63663, 2018.
7. S. A. Alabady, F. Al-Turjman, and S. Din, A novel security model for cooperative virtual networks in the IoT era, *International Journal of Parallel Programming*, 2018. doi:10.1007/s10766-018-0580-z.
8. F. Al-Turjman and S. Alturjman, Confidential smart-sensing framework in the IoT era, *The Journal of Supercomputing*, vol. 74, 5187–5198, 2018.
9. F. Al-Turjman and S. Alturjman, Context-sensitive access in industrial internet of things (IIoT) healthcare applications, *IEEE Transactions on Industrial Informatics*, vol. 14, no. 6, 2736–2744, 2018.
10. J. M. Hamamreh, H. M. Furqan, and H. Arslan, Classifications and applications of physical layer security techniques for confidentiality: A comprehensive survey, *IEEE Communications Surveys Tutorials*, 1, 2018. doi:10.1109/COMST.2018.2878035.
11. H. M. Furqan, J. M. Hamamreh, and H. Arslan, Secret key generation using channel quantization with SVD for reciprocal MIMO channels, in *International Symposium on Wireless Communication Systems* IEEE, Poznan, Poland, 2016, pp. 597–602.
12. J. M. Hamamreh, H. M. Furqan, and H. Arslan, Secure pre-coding and post-coding for OFDM systems along with hardware implementation, in *13th International Wireless Communications and Mobile Computing Conference, IWCMC 2017*, Valencia, Spain, June, 26–30, 2017, 2017, pp. 1338–1343.
13. H. M. Furqan, J. M. Hamamreh, and H. Arslan, Enhancing physical layer security of OFDM systems using channel shortening, in *Annual IEEE International Symposium on Personal, Indoor and Mobile Radio Communications, PIMRC 2017, Montreal, Canada*, October, 8–13, 2017, pp. 100–105.
14. E. Guvenkaya, J. M. Hamamreh, and H. Arslan, On physical-layer concepts and metrics in secure signal transmission, *Physical Communication*, vol. 25, 14–25, August 2017.
15. J. M. Hamamreh, E. Basar, and H. Arslan, OFDM-subcarrier index selection for enhancing security and reliability of 5G URLLC services, *IEEE Access*, vol. 5, 25863–25875, 2017.
16. J. M. Hamamreh and H. Arslan, Secure orthogonal transform division multiplexing (OTDM) waveform for 5G and beyond, *IEEE Communications Letters*, vol. 21, no. 5, 1191–1194, May 2017.

17. J. M. Hamamreh and H. Arslan, Joint PHY/MAC layer security design using ARQ with MRC and null-space independent PAPR-aware artificial noise in SISO systems, *IEEE Transactions on Wireless Communications*, vol. 17, no. 9, 6190–6204, September 2018.

18. Q. Wang, Z. Chen, W. Mei, and J. Fang, Improving physical layer security using UAV-enabled mobile relaying, *IEEE Wireless Communications Letters*, vol. 6, no. 3, 310–313, June 2017.

19. Q. Wang, Z. Chen, H. Li, and S. Li, Joint power and trajectory design for physical-layer secrecy in the UAV-aided mobile relaying system, *IEEE Access*, vol. 6, 62849–62855, 2018.

20. F. Cheng, G. Gui, N. Zhao, Y. Chen, J. Tang, and H. Sari, UAV relaying assisted secure transmission with caching, *IEEE Transactions on Communications*, 1, 2019. doi:10.1109/TCOMM.2019.2895088.

21. H. Liu, S. Yoo, and K. S. Kwak, Opportunistic relaying for low-altitude UAV swarm secure communications with multiple eavesdroppers, *Journal of Communications and Networks*, vol. 20, no. 5, 496–508, October 2018.

22. G. Zhang, Q. Wu, M. Cui, and R. Zhang, Securing UAV communications via joint trajectory and power control, *IEEE Transactions on Wireless Communications*, vol. 18, no. 2, 1376–1389, 2019.

23. M. Cui, G. Zhang, Q. Wu, and D. W. K. Ng, Robust trajectory and transmit power design for secure UAV communications, *IEEE Transactions on Vehicular Technology*, vol. 67, no. 9, 9042–9046, September 2018.

24. G. Zhang, Q. Wu, M. Cui, and R. Zhang, Securing UAV communications via joint trajectory and power control, *IEEE Transactions on Wireless Communications*, vol. 18, no. 2, 1376–1389, 2019.

25. Y. Zhu, G. Zheng, and M. Fitch, Secrecy rate analysis of UAV-enabled mmWave networks using matérn hardcore point processes, *IEEE Journal on Selected Areas in Communications*, vol. 36, no. 7, 1397–1409, July 2018.

26. Y. Zhou, P. L. Yeoh, H. Chen, Y. Li, W. Hardjawana, and B. Vucetic, Secrecy outage probability and jamming coverage of UAV-enabled friendly jammer, in *2017 11th International Conference on Signal Processing and Communication Systems (ICSPCS)*, Surfers Paradise, QLD, Australia, December 2017, pp. 1–6.

27. A. Li, Q. Wu, and R. Zhang, UAV-enabled cooperative jamming for improving secrecy of ground wiretap channel, *IEEE Wireless Communications Letters*, vol. 8, 181–184, 2018.

28. P. Nguyen, H. Nguyen, V.-D. Nguyen, and O.-S. Shin, UAV-enabled jamming noise for achieving secure communications in cognitive radio networks, 11 2018.

29. F. Shu, Z. Lu, J. Lin, L. Sun, X. Zhou, T. Liu, S. Zhang, W. Cai, J. Lu, and J. Wang, Alternating iterative secure structure between beamforming and power allocation for UAV-aided directional modulation networks, *Physical Communication*, vol. 33, 46–53, 2019.

30. A. Li and W. Zhang, Mobile jammer-aided secure UAV communications via trajectory design and power control, *China Communications*, vol. 15, no. 8, 141–151, August 2018.

31. H. Lee, S. Eom, J. Park, and I. Lee, UAV-aided secure communications with cooperative jamming, *IEEE Transactions on Vehicular Technology*, vol. 67, no. 10, 9385–9392, October 2018.

32. C. Zhong, J. Yao, and J. Xu, Secure UAV communication with cooperative jamming and trajectory control, *IEEE Communications Letters*, vol. 23, 286–289, 2019.

Index